Lecture Notes in Control and Information Sciences

Edited by A.V. Balakrishnan and M. Thoma

5

David J. Clements
Brian D. O. Anderson

Singular Optimal Control:
The Linear-Quadratic Problem

Springer-Verlag
Berlin Heidelberg New York 1978

Authors

David J. Clements,
School of Electrical Engineering,
University of New South Wales,
Kensington, N.S.W., 2033. Australia

Brian D. O. Anderson,
Department of Electrical Engineering,
University of Newcastle,
New South Wales, 2308, Australia.

ISBN 3-540-08694-3 Springer-Verlag Berlin Heidelberg New York
ISBN 0-387-08694-3 Springer-Verlag New York Heidelberg Berlin

Printing and binding: Beltz Offsetdruck, Hemsbach/Bergstr.
2061/3020-543210

PREFACE

Singular Optimal Control: The Linear-Quadratic Problem is a monograph aimed at
vanced graduate students, researchers and users of singular optimal control methods.
presumes prior exposure to the standard linear-quadratic regulator problem, and a
neral maturity in linear systems theory.

A number of advances in singular, linear-quadratic control have taken place very
cently. The book is intended to present an up-to-date account of many of these
vances. At the same time, the book attempts to present a unified view of various
proaches to singular optimal control, many of which are apparently unrelated.

knowledgements

The research reported   his book was supported by the Australian Research
ants Committee, and our   nks for this support are gratefully extended.

The manuscript typing, from first draft to final version, was undertaken by
s. Dianne Piefke. The willing and expert participation by Mrs. Piefke constituted
vital link in the publication chain, and to her the authors offer their sincerest
anks.

TABLE OF CONTENTS

CHAPTER I

SINGULAR LINEAR-QUADRATIC OPTIMAL CONTROL
- A BROAD BRUSH PERSPECTIVE

PROBLEM ORIGINS

Our concern throughout this book is with singular, linear-quadratic optimal con-
ol problems. In this section, we explore the origins of such problems, and in later
ctions of the chapter, we sketch some historical aspects, and describe how this book
rveys some major aspects of the present state of knowledge.

We first review the notion of a linear-quadratic problem (without regard to whe-
er or not it is singular), then we review the notion of a singular control problem
ithout regard to whether or not it is linear-quadratic), and then we tie the two
tions together.

Linear-quadratic optimal control problems, singular or nonsingular, usually arise
one of two distinct ways. First, there is prescribed a linear system

$$\dot{x} = F(t)x + G(t)u \quad x(t_0) = x_0 \quad \text{known} \tag{1.1}$$

d a performance index quadratic in u and x; for the linear regulation problem,
r example,

$$V[x_0, u(\cdot)] = \int_{t_0}^{t_f} [x'Q(t)x + u'R(t)u]dt + x'(T)Sx(T) \tag{1.2}$$

which, usually, Q, R and S are symmetric with Q and S nonnegative definite
d R positive definite. Of course, the problem is to find a control $u(\cdot)$ minimiz-
g the value of $V[x_0, u(\cdot)]$. The pioneering work of Kalman, see e.g. [1, 2] has
ven rise to an extensive study of this type of problem, see e.g. [3, 4]. Any perform-
ce index usually reflects some physically based notion of performance or quality of
ntrol, and (1.2) as a result of the listed constraints on Q, R and S is often
ry much physically based. However, it is possible to relax some of the constraints
Q, R and S and to allow a crossproduct term, $2x'H(t)u$ in the integrand of
.2). In this way, the most general form of linear-quadratic control problem can be
countered.

The second way in which linear-quadratic problems arise is via a perturbational
pe of analysis (a second variation theory) of a general optimal control problem, in
ich the underlying system may not be linear and the underlying performance index not
adratic. Given a certain initial state and the corresponding optimal control for a
neral optimal control problem, one can seek the adjustment to the optimal control
cessary to preserve optimality when the initial state is changed by a small amount;
approximation to the control adjustment follows as the solution to a linear-quadratic
oblem. For an exposition of the perturbation procedures including details of the cal-

culations for obtaining the linear-quadratic problem from the general problem, see e.g. [5].

Singular optimal control problems arise in the following way. With H the Hamiltonian, recall that, in any optimal control problem, singular or nonsingular, extremal arcs are defined by the requirement that H takes an extreme value. If this requirement does not allow the expression of the control vector in terms of state and costate vectors, the problem is singular. This can happen if H_u vanishes and H_{uu} is singular.

Johnson and Gibson demonstrated the existence of optimal singular solutions to certain problems in [6]; other examples are provided in [5, 7-9] and the excellent survey [10]. In most singular problems, the Hamiltonian is linear in $u(\cdot)$; this will be the case if the system equation and loss function involve $u(\cdot)$ linearly.

The notion of a singular linear-quadratic problem is obtained by a straightforward coalescing of the singular and linear-quadratic notions. Singularity of a linear-quadratic problem is equivalent to singularity of the matrix R in the cost function $x'Qx + 2u'Hx + u'Ru$. Singular linear-quadratic problems can arise directly, or as a result of applying a second variation theory to a general optimal control problem.

In considering any optimal control problem, a number of questions tend to arise. Is the problem solvable, or can the performance index be made as negative as desired by some choice of control? If the problem has a solution, how may one compute the optimal value of the performance index and an optimal control?

In the case of nonsingular linear-quadratic problems, as most readers will know, there are tidy solutions to these problems, see e.g. [5, 11-14]. Most of these solutions involve a matrix Riccati differential equation in which R^{-1} appears. As the survey [10] indicates, tidy solutions for singular problems in the main are much more recent or not yet available. We devote the next section to discussing some of the singular linear-quadratic results that have been determined.

2. HISTORICAL ASPECTS OF SINGULAR LINEAR-QUADRATIC CONTROL

Up till this point, at least four, largely disjoint, methods of attack on singular linear-quadratic problems have existed. In this section, we indicate what these are.

One thrust can be identified in the work of Goh [15, 16], Kelley [9, 17, 18] and Robbins [19]; this thrust may not have yet petered out, since further results are stil forthcoming, [20]. The prime, but not exclusive, concern in this work is with the computation problem, and the recurring theme in the papers is to aim to replace a singular linear-quadratic control problem by a nonsingular one, such that solutions of the nonsingular problem somehow determine solutions of the singular problem.

The second thrust is exemplified by the work of Jacobson [21, 22], as amplified by Anderson [23] and rounded nicely by Molinari [24]. Here the emphasis is on finding necessary and (sometimes differing) sufficient conditions for the solvability of sing-

r linear-quadratic problems. To be sure, conditions of one sort and another have
n known for a long time (e.g. $R \geq 0$); the point about the conditions in [21-24]
that they encompass all earlier known conditions, and when necessary and sufficient
dition sets differ, the difference is quite clearly very minor. Actually, very
ent work [25, 26] has been concerned with eliminating the differences.

The third method of attack is via regularization, i.e. one perturbs a singular
blem to make it nonsingular, the perturbation being such that the solution of the
singular problem is, in some sense, close to that of the singular problem. This
a has been exploited especially by Jacobson, see e.g. [27, 28], with the latter of
se references making contact with the problem of generating necessary and sufficient
ditions for problem solvability. The precise regularization procedure used is a
ple one - a term $\varepsilon u'u$ for small positive ε is added to the performance index
egrand. The effect is to perturb the optimal performance index slightly; however,
perturbation in the optimal control can be exceedingly hard to pin down, see e.g.
].

We shall pay very little attention to the regularization idea in this book, not
ause of any inherent demerit of the idea, but rather because once a nonsingular
blem has been obtained, the problem ceases to have much challenge about it. Further
se theoretical conditions concerning singular problems obtainable by studying a
uence of regularized problems with $\varepsilon \to 0$ are in the main obtainable more simply
other procedures. [Of course, in a specific problem, the study of such a sequence
deduce the optimal control may be very attractive computationally].

The fourth thrust is rather a related direction of research than a method of
ack on singular linear-quadratic problems. It turns out that there are certain
blems in passive network synthesis and covariance factorization that are allied to
singular linear-quadratic control problem. More precisely, certain matrix differ-
ial and integral inequalities recently perceived to be relevant in studying control
blems have been used for the network and covariance problems [30, 31].

By way of general comment, we note that experience shows that vector control
blems are often much more difficult than problems with a scalar control. Throughout
book, we consider vector controls.

OBJECTIVE OF THIS BOOK

We have set ourselves the task of presenting solutions to the existence and com-
ation questions associated with singular linear-quadratic problems. In the process
ever, we aim to show how the four directions of activity described in the last
tion may be made to coalesce, allowing the presentation of a unified theory.

Having done this, we attempt to translate a number of the ideas to discrete-time
blems. More is said about this in the next section.

4.   CHAPTER OUTLINE

We warn the reader that only brief comments on the background of the problems studied in this book are made in this section; we defer detailed comments to the introductory material in each chapter.

The central theme of Chapter II is robustness in linear-quadratic minimization problems. To understand why this should be so, we shall digress from a description of the actual chapter contents.

In studying the second variation problem, it is frequently the case that the initial state in the associated linear-quadratic problem is always zero, and one desires necessary and sufficient conditions for the performance index to be nonnegative for all controls $u(\cdot)$. With a controllability assumption, a necessity condition can be stated, and without the controllability assumption, a sufficiency condition can be found, [21-24]. The necessity and sufficiency conditions are very similar but not identical; the sufficiency condition is basically a generalization to the singular case of a condition that a Riccati equation appearing in the nonsingular problem have no escape times on $[t_0, t_f]$.

The question then arises as to how this aesthetically disquieting situation of differing necessary and sufficient conditions for the nonnegativity requirement can be remedied. The solution to the problem is quite simple – we slightly change the problem statement. Instead of seeking necessary and sufficient conditions for nonnegativity we seek necessary and sufficient conditions for the optimal performance index to be finite for all initial states. Equivalently, we demand that the minimization problem (or, more generally, the infimization problem) have a solution not just for zero initial state, but for all initial states close to zero (and hence for all initial states without restriction on size, by the linear-quadratic nature of the problem). We suggest that any realistic model of a physical control system that has a solution for zero initial condition should also have a solution for any initial condition in a neighbourhood of zero. Otherwise for an arbitrarily small change in initial condition of the system the optimal cost in controlling the system would change from a finite value to $-\infty$. Clearly, such a situation is unrealistic.

The first use of the robustness idea in Chapter II is therefore, in the relatively easy derivation of identical necessary and sufficient conditions for the linear-quadratic problem to have a finite optimal performance index for all initial states. Once having introduced the idea of robustness however, we are led to examine the extent of its applicability to other types of robustness, e.g. with respect to initial time, final time, end-point weighting, and initial and end-point constraints.

Actually, the notion of robustness with respect to initial time has been used elsewhere, as in [14] where it is termed "extendability". It turns out that a minimization problem is robust with respect to the initial time if and only if it is robust with respect to the initial state.

The latter part of Chapter II is concerned with constrained end-point problems.

e we study the existence question again, once more closing gaps between earlier
essity and sufficiency conditions by imposing a robustness assumption. It turns
that this time there are three possible robustness assumptions, not two, which
nontrivially equivalent: robustness with respect to final state, final time and
minal state weighting matrix in the performance index.

Throughout Chapter II, conditions are almost entirely expressed in terms of matrix
tegral inequalities, of the type that have evolved from the second general line of
ack on singular problems described earlier [21-24]. The opportunity is therefore
ken to derive several new properties of these inequalities. Some have appeared in
, 26], while others are reported here for the first time.

Whereas Chapter II is devoted to the existence question, Chapter III is concerned
th computation questions - more precisely with describing algorithms for checking
negativity of the performance index with zero initial state and arbitrary control,
evaluating the optimal performance index with arbitrary initial state (and in the
cess checking its existence), and/or computing optimal controls. The algorithms
shown to flow from the first, second and fourth approaches to singular problems
out earlier, i.e. all approaches save that of regularization. This provides
erefore a significant unification of many ideas previously somewhat disjoint.

There is, unfortunately, a caveat. To execute the algorithms, certain smoothness
constancy of rank assumptions must be fulfilled, and this cannot always be guaran-
d.

While we obviously leave precise specification of the algorithms to Chapter III,
mention several global aspects of them here. The algorithms proceed by replacing
singular linear-quadratic problem by another linear-quadratic problem with either
ver control space dimension and/or lower state space dimension. The replacing
oblem may be nonsingular or singular. A series of such replacements ultimately
ads to a problem which either has zero control space dimension, zero state space
nension, or is nonsingular. In all three cases, the problem is easily dealt with,
its solution can be somehow translated back to provide a solution to the origin-
y posed singular problem.

In Chapter IV, singular discrete-time problems are defined and discussed. From
point of view, one might claim there is little to be gained from a close examination
discrete-time problems, since a Riccati difference equation defining the optimal cost
a formula for the optimal control are always valid. Nevertheless, many interesting
allels with the singular continuous-time results can be obtained, including a very
eral theory of degenerate and constant directions, first introduced and explored in
-34].

Chapter V offers a very brief statement of some remaining directions for research.

ERENCES

R.E. Kalman, "Contributions to the theory of optimal control", *Bol. Soc. Mat. Mex.*
1960, pp. 102-119.

[2] R.E. Kalman, "The Theory of Optimal Control and the Calculus of Variations", Chapter 16 of Mathematical Optimization Techniques, ed. R. Bellman, University of California Press, 1963.

[3] H. Kwakernaak and R. Sivan, Linear Optimal Control Systems, Wiley-Interscience, New York, 1972.

[4] B.D.O. Anderson and J.B. Moore, Linear Optimal Control, Prentice-Hall, New Jersey, 1971.

[5] A.E. Bryson and Y.C. Ho, Applied Optimal Control, Blaisdell Publishing Co., Mass., 1969.

[6] C.D. Johnson and J.E. Gibson, "Singular solutions in problems of optimal control", IEEE Trans. Automatic Control, Vol. AC-8, 1963, pp. 4-15.

[7] W.M. Wonham and C.D. Johnson, "Optimal bang-bang control with quadratic performance index", Trans. ASME, Series D, J. Bas. Eng., Vol. 86, 1964, pp. 107-115.

[8] C.D. Johnson and W.M. Wonham, "On a problem of Letov in optimal control", Trans. ASME, Series D, J. Bas. Eng., Vol. 87, 1965, pp. 81-89.

[9] H.J. Kelley, R.E. Kopp and H.G. Moyer, "Singular Extremals", Chapter 3 in Topics in Optimization, ed. by G. Leitmann, Academic Press, New York, 1967.

[10] D.A. Bell and D.H. Jacobson, Singular Optimal Control Problems, Academic Press, New York, 1975.

[11] I.M. Gelfand and S.W. Fomin, Calculus of Variations, Prentice-Hall, New Jersey, 1963.

[12] J.V. Breakwell, J.L. Speyer and A.E. Bryson, "Optimization and control of nonlinear systems using the second variation", SIAM J. Control, Vol. 1, 1963, pp. 193-223.

[13] W.A. Coppel, Disconjugacy, Lecture Notes in Mathematics, Vol. 220, Springer-Verlag, Berlin, 1971.

[14] J.B. Moore and B.D.O. Anderson, "Extensions of quadratic minimization theory, I: Finite time results", Int. J. Control, Vol. 7, 1968, pp. 465-472.

[15] B.S. Goh, "The second variation for the singular Bolza problem", SIAM J. Control, Vol. 4, 1966, pp. 309-325.

[16] B.S. Goh, "Necessary conditions for singular extremals involving multiple control variables", SIAM J. Control, Vol. 4, 1966, pp. 716-731.

[17] H.J. Kelley, "A second variation test for singular extremals", AIAA J., Vol. 2, 1964, pp. 1380-1382.

[18] H.J. Kelley, "A transformation approach to singular subarcs in optimal trajectory and control problems", SIAM J. Control, Vol. 2, 1964, pp. 234-240.

[19] H.M. Robbins, "A generalized Legendre-Clebsch condition for the singular cases of optimal control", IBM J. Res. Develop., Vol. 3, 1967, pp. 361-372.

[20] J.B. Moore, "The singular solutions to a singular quadratic minimization problem", Int. J. Control, Vol. 20, 1974, pp. 383-393.

[21] D.H. Jacobson, "Totally singular quadratic minimization problems", IEEE Trans. Automatic Control, Vol. AC-16, 1971, pp. 651-658.

2] J.L. Speyer and D.H. Jacobson, "Necessary and sufficient conditions for optimality for singular control problems: a transformation approach", *J. Math. Anal. Appl.*, Vol. 33, 1971, pp. 163-187.

3] B.D.O. Anderson, "Partially singular linear-quadratic control problems", *IEEE Trans. Automatic Control*, Vol. AC-18, 1973, pp. 407-409.

4] B.P. Molinari, "Nonnegativity of a quadratic functional", *SIAM J. Control*, Vol. 13, 1975, pp. 792-806.

5] D.J. Clements, B.D.O. Anderson and P.J. Moylan, "Matrix inequality solution to linear-quadratic singular control problems", *IEEE Trans. Automatic Control*, Vol. AC-22, 1977, pp. 55-57.

6] D.J. Clements and B.D.O. Anderson, "Transformational solution of singular linear-quadratic control problems", *IEEE Trans. Automatic Control*, Vol. AC-22, 1977, pp. 57-60.

7] D.H. Jacobson, S.B. Gershwin and M.M. Lele, "Computation of optimal singular controls", *IEEE Trans. Automatic Control*, Vol. AC-15, No. 1, February 1970, pp. 67-73.

8] D.H. Jacobson and J.L. Speyer, "Necessary and sufficient conditions for singular control problems: a limit approach", *J. Math. Anal. Appl.*, Vol. 34, 1971, pp.239-266.

9] R.E. O'Malley, Jr., and A. Jameson, "Singular perturbations and singular arcs - Part I", *IEEE Trans. Automatic Control*, Vol. AC-20, 1975, pp. 218-226.

0] B.D.O. Anderson and P.J. Moylan, "Synthesis of linear time varying passive networks", *IEEE Trans. Circuits and Systems*, Vol. CAS-21, 1974, pp. 678-687.

1] B.D.O. Anderson and P.J. Moylan, "Spectral factorization of a finite-dimensional nonstationary matrix covariance", *IEEE Trans. Automatic Control*, Vol. AC-19, 1974, pp. 680-692.

2] R.S. Bucy, D. Rappaport and L.M. Silverman, "Correlated noise filtering and invariant directions of the Riccati equation", *IEEE Trans. Automatic Control*, Vol. AC-15, 1970, pp. 535-540.

3] D. Rappaport, "Constant directions of the Riccati equation", *Automatica*, Vol. 8, 1972, pp. 175-186.

4] M. Gevers and T. Kailath, "Constant, predictable and degenerate directions of the discrete-time Riccati equation", *Automatica*, Vol. 9, 1973, pp. 699-711.

CHAPTER II

ROBUST LINEAR-QUADRATIC MINIMIZATION

1.   INTRODUCTION

Conditions for the solvability of linear-quadratic continuous-time minimization
problems have been studied in a number of papers, e.g. [1-7]. The more recent work has
shown the value of characterizing problem solvability in terms of nonnegativity conditions involving certain Riemann-Stieltjes integrals. In this work, one somewhat
bothersome problem remains - there are generally gaps between necessary conditions and
sufficient conditions. In this chapter, we survey this material and show how to eliminate these gaps.

In the broadest terms, what we find is that the gap vanishes if one is studying
problems which are in some way robust, that is, tolerant of small changes to some or
all of the relevant parameters. To the extent that nonrobust problems can be regarded
as qualitatively ill-posed (though in strict terms, or quantitative terms, they may be
well-posed), we are saying that there is no gap in the case of a well-posed problem.

In what ways do we vary parameters in the robustness study? The answer depends
somewhat on the problem in hand. For free end-point problems, we can consider a variation in the initial time, and in the case of these conditions applying to a zero
initial state situation, we can consider a variation away from zero of that initial
state. For constrained end-point problems, we can consider a variation in the final
time, we can consider the use of a penalty function approach, and we can consider a
relaxation of a constraint requiring that the final state be at a point to one requiring the final state to be in a (small) sphere around that point. For both classes of
problems, we can also consider variations in the parameters of the matrices defining
the system and performance index. (We do not however explore this last issue deeply;
one immediate and major difficulty stems from deciding which quantities normally zero
should remain at zero in a general parameter change, and which should not). Altogether
then, in this chapter we study a range of robustness problems. In general, many of our
conclusions are along the lines that if there is one kind of robustness, another kind
is automatically implied.

An outline of this chapter is as follows. In Section 2, we study an arbitrary
linear-quadratic problem with arbitrary constraints on the initial and terminal states.
We show that if the constraints are achievable, the optimal performance index is a
quadratic form in the independent variables which define the initial and terminal states
Various versions of this result are used through the later sections. In Section 3, we
study problems with arbitrary but fixed initial state and free terminal state. Our
main results equate two types of robustness (with respect to initial state and initial
time) with an inequality condition involving Riemann-Stieltjes integrals. In Section
4, we turn to constrained end-point problems; the use of penalty functions provides
yet another form of robustness, and most of the work is concerned with showing the

uivalence of this new type of robustness with robustness properties involving the

rminal time or state, and with a Riemann-Stieltjes integral inequality condition.

nally in Section 5 we study the extremal solutions of the Riemann-Stieltjes inequ-

ity of Sections 3 and 4 and the relation between these extremal solutions and optim-

ation problems. Section 6 offers some summarizing remarks.

QUADRATIC PROPERTY OF THE OPTIMAL PERFORMANCE INDEX

We study the system

$$\dot{x} = F(t)x + G(t)u \qquad\qquad t_0 \le t \le t_f \qquad\qquad\qquad (2.1)$$

th various possible constraints on the initial and terminal states. Thus we post-

ate that for two fixed (possibly evanescing) constant matrices D_0, E_0, we have

$$D_0 x(t_0) = \eta_0 \qquad\qquad E_0 x(t_0) = 0, \qquad\qquad\qquad (2.2)$$

d that the controls are constrained such that

$$D_f x(t_f) = \eta_f \qquad\qquad E_f x(t_f) = 0. \qquad\qquad\qquad (2.3)$$

(2.2) and (2.3), D_0, E_0, D_f and E_f are fixed constant matrices such that (with-

it loss of generality) $C_0 = [D_0' \ E_0']'$ and $C_f = [D_f' \ E_f']'$ have full row rank. We

low the possibility of one or more of D_0, E_0, etc. evanescing. The vectors η_0

d η_f are arbitrary, but fixed in any given problem. The commonest situations are

ose where $D_0 = I$, E_0, D_f and E_f evanescent (the usual free terminal state prob-

em) and $D_0 = I$, E_0 and D_f evanescing, and $E_f = I$ (the usual zero terminal state

oblem).

Associated with (2.1) through (2.3) is the performance index

$$V[x_0, t_0, u(\cdot)] = \int_{t_0}^{t_f} \{x'Q(t)x + 2u'H(t)x + u'R(t)u\}dt$$

$$+ x'(t_f)Sx(t_f) \qquad\qquad (2.4)$$

here $x_0 = x(t_0)$. The matrices $F(\cdot)$, $G(\cdot)$, $H(\cdot)$, $Q(\cdot)$ and $R(\cdot)$ all have dimen-

ions consistent with (2.1) and (2.4) and are piecewise continuous. (This has the

bvious meaning; the matrices have piecewise continuous entries). The matrix S is

onstant. Without loss of generality, we assume $Q(\cdot)$, $R(\cdot)$ and S are symmetric.

he controls $u(\cdot)$ are assumed to be piecewise continuous on $[t_0, t_f]$. To begin

ith, we shall study classes of problems in which one or both of x_0 and x_f vary;

ater, variation of t_0 and t_f will be considered. In general, we shall be inter-

sted in the question of when (2.4) possesses an infimum which is finite.

However, in this section, we shall be concerned with establishing the functional

orm of the infimum. For each x_0 and $u(\cdot)$, (2.4) takes a definite value. Now

magine that η_0, η_f are fixed but arbitrary, and that x_0 and $u(\cdot)$ exist such that (2.1) through (2.3) are satisfied. Then essentially (2.4) is being evaluated when x_0 and $u(\cdot)$ are constrained linearly, by (2.1) through (2.3). In general, each η_0, η_f pair will permit an infinity of x_0, $u(\cdot)$ satisfying (2.1) through (2.3). We define $V^*[\eta_0, \eta_f]$ to be the infimum of the values of $V[x_0, t_0, u(\cdot)]$ obtainable under the constraints (2.1) through (2.3). In case $D_0 = I$, and E_0, D_f and E_f, evanesce, then $V^*[\eta_0, \eta_f]$ simply becomes $V^*[x_0]$, the usual free-end-point optimal performance index.

The main idea of this section is that the <u>optimal performance index, if it is finite for all η_0 and η_f, is quadratic in these quantities.</u> Special cases of the result will be used in later sections. We also postpone to later sections study of issues of existence and computation.

Before proving the quadratic property however, we consider a preliminary question: for a given η_0, η_f pair, is it always possible to find $x(t_0)$ and $u(\cdot)$ satisfying (2.1) through (2.3)? Various special answers are easily established. Thus if D_0 and E_0 evanesce, i.e. $x(t_0)$ is free, then there always exists an $x(t_0)$ and $u(\cdot)$ such that $D_f x(t_f) = \eta_f$, $E_f x(t_f) = 0$ for any η_f. (Because $[D_f'\ E_f']'$ has full row rank, there is no possibility of the constraints on $x(t_f)$ being inconsistant). On the other hand, if $D_0 = I$, E_0, and D_f evanesce and $E_f = I$, so that controls are required to take an arbitrary $x(t_0) = \eta_0$ to $x(t_f) = 0$, it is clear that a controllability condition of some description is required. The precise condition is pinned down in the following way.

<u>Lemma II.3.1.</u> Let $V^*[\eta_0, \eta_f]$ denote inf $V[x(t_0), t_0, u(\cdot)]$ [defined in (2.4)], subject to (2.1) through (2.3) holding. Then $V^*[\eta_0, \eta_f] < \infty$ for all n_0, η_f [equivalently: (2.1) through (2.3) are attainable for arbitrary η_0, η_f] if and only if

$$\text{Range}\left\{\begin{bmatrix} I \\ 0 \end{bmatrix} \vdots\ C_f \Phi(t_f,\ t_0) C_0'(C_0 C_0')^{-1}\begin{bmatrix} I \\ 0 \end{bmatrix}\right\}$$

$$\subset \text{Range }\{C_f[\Phi(t_f,\ t_0)K \vdots\ W]\}$$

(2.5)

where K is a matrix whose columns constitute a basis for $N(C_0)$, the nullspace of C_0, and W is the controllability matrix

$$W = \int_{t_0}^{t_f}\Phi(t_f,\ \tau)G(\tau)G'(\tau)\Phi'(t_f,\ \tau)d\tau.$$

<u>Proof</u>: Necessity: observe that for any $u(\cdot)$, $\int_{t_0}^{t_f}\Phi(t_f,\ \tau)G(\tau)u(\tau)d\tau$ has the form $W\alpha$ for some α [8, see p. 75], and that any $x(t_0)$ for which $C_0 x(t_0) = [\eta_0'\ 0]'$ may be written as $C_0'(C_0 C_0')^{-1}[\eta_0'\ 0]' + K\beta$ for some β. Now premultiply by C_f the basic equation

$$x(t_f) = \Phi(t_f,\ t_0)x(t_0) + \int_{t_0}^{t_f}\Phi(t_f,\ \tau)G(\tau)u(\tau)d\tau.$$

This leads to

$$\begin{bmatrix} \eta_f \\ 0 \end{bmatrix} - C_f \Phi(t_f, t_0) C_0'(C_0 C_0')^{-1} \begin{bmatrix} \eta_0 \\ 0 \end{bmatrix} = C_f \Phi(t_f, t_0) K \beta + C_f W \alpha \qquad (2.6)$$

some α and β. Now in order that (2.3) be attainable given (2.1) and (2.2) there
be α and β satisfying (2.6). In order that such α and β exist for all
and η_f, (2.5) must hold.

Sufficiency: for prescribed η_f, η_0, choose α, β satisfying (2.6), which is
ible in view of (2.5). Take $u(t) = G'(t) \Phi'(t_f, t) \alpha$ and $x(t_0) = C_0'(C_0 C_0')^{-1} [\eta_0' \quad 0]'$.
(2.2) and (2.3) both hold. This proves the lemma.

rks 2.1: 1. In case $x(t_0)$ is fixed but arbitrary i.e. $D_0 = I$, while E_0 and
evanesce, the condition of the lemma is equivalent to the positive definiteness of
E_f', which is used in [4]. Whatever values D_0, etc., take, positive definite-
of $C_f W C_f'$ certainly causes satisfaction of (2.5).

    2. The question of whether or not $V^*[\eta_0, \eta_f] > -\infty$ is entirely different
whether or not $V^*[\eta_0, \eta_f] < \infty$, and will be discussed in later sections. (It is
od deal harder to answer).

To establish the quadratic nature of $V^*[\eta_0, \eta_f]$, we shall use the following
acterization of quadratic functions. A similar characterization has been used
where in studying linear-quadratic problems [4, 9].

Lemma II.2.2. Let $K(\cdot)$ be a scalar function of an n-vector. Then $K(\cdot)$ is
quadratic, i.e. $K(x) = x'Px$ for some symmetric P if and only if for all scalar
λ and n-vector x_1, x_2

$$K(\lambda x_1) = \lambda^2 K(x_1)$$

$$K(x_1 + x_2) + K(x_1 - x_2) = 2K(x_1) + 2K(x_2) \qquad (2.7)$$

$$K(x_1 + \lambda x_2) - K(x_1 - \lambda x_2) = \lambda K(x_1 + x_2) - \lambda K(x_1 - x_2).$$

f: The "only if" part of the result is easily checked. For the "if" part, we
eed as follows. Below, it is shown that (2.7) imply

$$K(x_1 + x_2) - K(x_1 - x_2) + K(x_1 + x_3) - K(x_1 - x_3)$$

$$= K(x_1 + x_2 + x_3) - K(x_1 - x_2 - x_3). \qquad (2.8)$$

ming this for the moment, set

$$L(x, y) = K(x+y) - K(x-y).$$

(2.8) shows that $L(x, y_1 + y_2) = L(x, y_1) + L(x, y_2)$, the first equation in (2.7)
ds that $L(x, y) = L(y, x)$ since $K(x-y) = K(y-x)$, and the third equation of (2.7)
ds that $L(x, \lambda y) = \lambda L(x, y)$. Therefore L is bilinear in x and y. Consequently,

$L(x, x)$ is quadratic and noting that $L(x, x) = K(2x) - K(0) = 4K(x)$ [using the first equation of (2.7)], we also have $K(x)$ quadratic.

We now verify (2.8). We have, from the second equation of (2.7),

$$K(x_1+x_2) + K(x_1+x_3) = \tfrac{1}{2}[K(x_2-x_3) + K(2x_1+x_2+x_3)]$$

$$K(x_1-x_2) + K(x_1-x_3) = \tfrac{1}{2}[K(2x_1-x_2-x_3) + K(x_2-x_3)]$$

whence

$$K(x_1+x_2) - K(x_1-x_2) + K(x_1+x_3) - K(x_1-x_3) + K(x_1-x_2-x_3) - K(x_1+x_2+x_3)$$

$$= \tfrac{1}{2}[K(2x_1+x_2+x_3) - 2K(x_1+x_2+x_3)] - \tfrac{1}{2}[K(2x_1-x_2-x_3) - 2K(x_1-x_2-x_3)].$$

The right hand side of this equation is identical to

$$\tfrac{1}{2}\{K(x_2+x_3+2x_1) - K(x_2+x_3-2x_1)$$

$$-2K(x_2+x_3+x_1) + 2K(x_2+x_3-x_1)\}$$

and so, using the third equation of (2.7) with $\lambda=2$, it becomes zero. Thus, (2.8) follows.

The quadratic nature of $V^*[\eta_0, \eta_f]$ can now be established.

Theorem II.2.1. Suppose $V^*[\eta_0, \eta_f]$ exists for all η_0 and η_f. Then $V^*[\eta_0, \eta_f]$ has the representation

$$V^*[\eta_0, \eta_f] = [\eta_0' \quad \eta_f'] \begin{bmatrix} P_{00} & P_{0f} \\ P_{0f} & P_{ff} \end{bmatrix} \begin{bmatrix} \eta_0 \\ \eta_f \end{bmatrix} \tag{2.9}$$

for some matrices P_{00}, P_{0f}, P_{ff}.

Proof: As a consequence of the above lemma, the desired result follows if $V^*[\eta_0, \eta_f]$ satisfies the three equalities

$$V^*[\lambda\eta_0, \lambda\eta_f] = \lambda^2 V^*[\eta_0, \eta_f] \tag{2.10}$$

$$V^*[\eta_{01}+\eta_{02}, \eta_{f1}+\eta_{f2}] + V^*[\eta_{01}-\eta_{02}, \eta_{f1}-\eta_{f2}]$$

$$= 2V^*[\eta_{01}, \eta_{f1}] + 2V^*[\eta_{02}, \eta_{f2}] \tag{2.11}$$

and

$$V^*[\eta_{01}+\lambda\eta_{02}, \eta_{f1}+\lambda\eta_{f2}] - V^*[\eta_{01}-\lambda\eta_{02}, \eta_{f1}-\lambda\eta_{f2}]$$

$$= \lambda V^*[\eta_{01}+\eta_{02}, \eta_{f1}+\eta_{f2}] - \lambda V^*[\eta_{01}-\eta_{02}, \eta_{f1}-\eta_{f2}]. \tag{2.12}$$

method of proof is very similar for each case and proceeds by contradiction; it
ies on the quadratic nature of $V[x(t_0), t_0, u(\cdot)]$ in (2.4), which implies versions
(2.10) through (2.12) for V as opposed to V^*. We shall prove merely (2.12), this
ng slightly harder than (2.10) and (2.11).

The third equality is trivial for $\lambda = \pm 1$ and zero, and if true for $\lambda > 0$ is
ily extended to $\lambda < 0$. So assume $\lambda > 0$, $\lambda \neq 1$.

Suppose that $u_i(\cdot)$, $i = 1, 2$ are any controls ensuring satisfaction of the
straints (2.1) through (2.3) for η_0, η_f replaced by η_{0i}, η_{fi} for $i = 1, 2$
pectively. By linearity of (2.1) through (2.3) $u_1 + \mu u_2$ will ensure satisfaction
h η_0, η_f replaced by $\eta_{01} + \mu\eta_{02}$, $\eta_{f1} + \mu\eta_{f2}$. From the quadratic nature of the
formance index (2.4), it is easily checked that

$$V[x_{01}+\lambda x_{02}, t_0, u_1+\lambda u_2] + \lambda V[x_{01}-x_{02}, t_0, u_1-u_2]$$

$$= \lambda V[x_{01}+x_{02}, t_0, u_1+ u_2] + V[x_{01}-\lambda x_{02}, t_0, u_1-\lambda u_2]. \qquad (2.13)$$

arbitrary $\varepsilon > 0$, choose $u_3(\cdot)$ and $u_4(\cdot)$ so that

$$V[x_{01}+x_{02}, t_0, u_3] \leq V^*[\eta_{01}+\eta_{02}, \eta_{f1}+\eta_{f2}] + \varepsilon$$

$$V[x_{01}-\lambda x_{02}, t_0, u_4] \leq V^*[\eta_{01}-\lambda\eta_{02}, \eta_{f1}-\lambda\eta_{f2}] + \varepsilon. \qquad (2.14)$$

o, u_3 is to cause satisfaction of (2.1) through (2.3) with η_0, η_f replaced by
$+\eta_{02}$, $\eta_{f1}+\eta_{f2}$ while u_4 is to cause satisfaction with replacement by $\eta_{01}-\lambda\eta_{02}$,
$-\lambda\eta_{f2}$. Define $u_1(\cdot)$, $u_2(\cdot)$ by

$$u_1+u_2 = u_3$$

$$u_1-\lambda u_2 = u_4 .$$

n (by linearity) $u_1+\lambda u_2$ and u_1-u_2 cause satisfaction of (2.1) through (2.3)
pairs $\eta_{01}+\lambda\eta_{02}$, $\eta_{f1}+\lambda\eta_{f2}$ and $\eta_{01}-\eta_{02}$, $\eta_{f1}-\eta_{f2}$ respectively. Equations (2.13)
(2.14) now yield

$$V^*[\eta_{01}+\lambda\eta_{02}, \eta_{f1}+\lambda\eta_{f2}] + \lambda V^*[\eta_{01}-\eta_{02}, \eta_{f1}-\eta_{f2}]$$

$$\leq V[x_{01}+\lambda x_{02}, t_0, u_1+\lambda u_2] + \lambda V[x_{01}-x_{02}, t_0, u_1-u_2]$$

$$\leq \lambda V^*[\eta_{01}+\eta_{02}, \eta_{f1}+\eta_{f2}] + V^*[\eta_{01}-\lambda\eta_{02}, \eta_{f1}-\lambda\eta_{f2}] + (1+\lambda)\varepsilon.$$

imilar argument yields an inequality going the other way and since ε is arbitrary,
obtains

$$V^*[\eta_{01}+\lambda\eta_{02}, \eta_{f1}+\lambda\eta_{f2}] + \lambda V^*[\eta_{01}-\eta_{02}, \eta_{f1}-\eta_{f2}]$$

$$= \lambda V^*[\eta_{01}+\eta_{02}, \eta_{f1}+\eta_{f2}] + V^*[\eta_{01}-\lambda\eta_{02}, \eta_{f1}-\lambda\eta_{f2}]$$

as required.

Remarks 2.2: In case $D_0 = I$ and E_0, D_f and E_f evanesce (the usual optimal control problem), a version of the above result is proved in [4] with an additional and substantial controllability constraint, over and above that of Lemma II.2.1 (which in this case evanseces). The first removal of this controllability constraint appears to be in [6].

3. INITIAL CONDITION RESULTS AND THE RIEMANN-STIELTJES INEQUALITY

In this section, we study the system

$$\dot{x} = F(t)x + G(t)u \qquad t_0 \leq t \leq t_f \qquad (3.1)$$

with $x(t_0) = x_0$ prescribed, $x(\cdot)$ of dimension n and $u(\cdot)$ of dimension m. Associated with (3.1) is the performance index

$$V[x_0, t_0, u(\cdot)] = \int_{t_0}^{t_f} \{x\acute{}Q(t)x + 2u\acute{}H(t)x + u\acute{}R(t)u\}dt$$
$$+ x\acute{}(t_f)Sx(t_f). \qquad (3.2)$$

The matrices $F(\cdot)$, $G(\cdot)$, $H(\cdot)$, $Q(\cdot)$ and $R(\cdot)$ all have dimensions consistent with (3.1) and (3.2) and are piecewise continuous. (This has the obvious meaning; the matrices have piecewise continuous entries). In the next chapter, stronger smoothness conditions will be imposed. The matrix S is constant. Without loss of generality, we assume $Q(\cdot)$, $R(\cdot)$ and S are symmetric. The controls $u(\cdot)$ are assumed to be piecewise continuous on $[t_0, t_f]$. We specialise the ideas of the last section, by assuming x_f is free; with x_0 arbitrary fixed, this is a standard optimal control problem. It is immediate that V, and thus V^*, cannot be $+\infty$. Here, we shall be interested in the question of when (2.2) possesses an infimum, denoted $V^*[x_0, t_0]$, which is not $-\infty$.

The problem of determining when (3.2) has a finite infimum subject to the constraint (3.1) is called nonsingular, partially singular or totally singular according as $R(t) > 0$ on $[t_0, t_f]$, $R(t)$ is singular at one point on $[t_0, t_f]$ but not identically zero, or $R(t) \equiv 0$ on $[t_0, t_f]$. Of course, a well-known necessary condition for a finite infimum is $R(\cdot) \geq 0$ on $[t_0, t_f]$, this being the classical Legendre-Clebsch necessary condition [10].

The nonsingular problem is much easier than the singular problem, which will take most of our attention. However, the earlier results of this section apply equally to nonsingular and singular problems.

We remark also that in this book we study various necessity and sufficiency conditions for $V^*[x_0, t_0]$ to be finite in reverse order to their historical development. In particular, in this chapter we are interested in general existence conditions for finiteness of $V^*[x_0, t_0]$ whereas in Chapter III, we turn our attention to a more

sical approach which, on the one hand, is restrictive in that not all problems
be covered, while on the other hand, lends itself to the derivation of computational
rithms for computing optimal controls and performance indices.

As noted above, we shall assume throughout this section that $x(t_f)$ is free,
ng up the possibility of constraining the value of $x(t_f)$ in later sections.
ver, the reader should be aware that all the results of this section carry over to
constrained end-point problem with minor changes, and these minor changes do <u>not</u>
lve the behaviour of quantities in the vicinity of t_0. Since the results of this
ion are almost all concerned with behaviour in the vicinity of t_0, these minor
ges are also conceptually insignificant.

We now define further notation. With $P(\cdot)$ an $n \times n$ symmetric matrix with entries
ounded variation, set

$$dM(P) = \begin{bmatrix} dP + (PF+F'P+Q)dt & (PG+H)dt \\ (PG+H)'dt & Rdt \end{bmatrix} . \tag{3.3}$$

subsequent material will make heavy use of the following inequality:

$$\int_{t_1}^{t_2} [v'(t) \quad u'(t)] \; dM(P) \begin{bmatrix} v(t) \\ u(t) \end{bmatrix} \geq 0 \tag{3.4}$$

any piecewise continuous m-vector $u(\cdot)$, and continuous n-vector $v(\cdot)$ defined
$[t_1, t_2]$. The integral is defined in the usual Riemann-Stieltjes sense. We shall
that $dM(P) \geq 0$ or $dM(P)$ is nonnegative within any interval I, closed or open
ither end, if the inequality holds on $[t_1, t_2]$ for all $[t_1, t_2] \subset I$.

With these preliminaries, we can indicate the principal known connections between
existence of finite infima for the minimization problem, and the existence of
ices P satisfying $dM(P) \geq 0$ on certain intervals and satisfying a terminal
ition.

A convenient reference for the first two theorems is [4]. However, Theorem II.3.1
extend the main result of [4] in a manner we explain subsequently.

<u>Theorem II.3.1</u>: Suppose that $V[0, t_0, u(\cdot)] \geq 0$ for all $u(\cdot)$ and that

$$\int_{t_0}^{t_1} \Phi(t_1, \tau)G(\tau)G'(\tau)\Phi'(t_1, \tau)d\tau > 0 \quad \forall \; t_1 \in (t_0, t_f] \tag{3.5}$$

(so that all states are reachable from $x(t_0) = 0$ at any time $t_1 > t_0$). Here,
$\Phi(\cdot, \cdot)$ is the transition matrix associated with $F(\cdot)$. Then

$$\infty > V^*[x_1, t_1] > -\infty \quad \forall \; x(t_1) = x_1, \quad \forall \; t_1 \in (t_0, t_f] \tag{3.6}$$

and there exists a symmetric $P(\cdot)$ of bounded variation defined on $(t_0, t_f]$
such that $P(t_f) \leq S$ and $dM(P) \geq 0$ within $(t_0, t_f]$.

The proof of this theorem will proceed by a series of lemmas. Broadly speaking, the strategy is as follows. We first demonstrate (3.6); then we appeal to the material of the last section to conclude a quadratic form for $V^*[x_1, t_1]$ as $x_1'P^*(t_1)x_1$ for some $P^*(t_1)$; we show that $P^*(\cdot)$ is of bounded variation and satisfies a restricted form of the requirement $dM(P^*) \geq 0$, and then finally we remove the restriction.

<u>Lemma II.3.1</u>: With the hypotheses of Theorem II.3.1, $V^*[x_1, t_1]$ is finite for all $x(t_1) = x_1$ and $t_1 \in (t_0, t_f]$.

<u>*Proof*</u>: The inequality $\infty > V^*[x_1, t_1]$ for each $x_1 = x(t_1)$ and each $t_1 \in (t_0, t_f]$ is trivial. For the other inequality $V^*[x_1, t_1] > -\infty$, we have by the reachability condition (3.5) that there exists a piecewise continuous control $u_r(t)$ on $[t_0, t_1]$ taking $x = 0$ at time t_0 to $x = x_1$ at time t_1. Then letting $u(t)$ be any piecewise continuous control on $[t_1, t_f]$, we have by assumption that

$$\int_{t_0}^{t_1} q(x_r, u_r)dt + \int_{t_1}^{t_f} q(x, u)dt + x'(t_f)Sx(t_f) \geq 0,$$

where $q(x, u) = x'Qx + 2x'Hu + u'Ru$. Clearly this gives a lower bound on $V[x_1, t_1, u(\cdot)]$ and hence $V^*[x_1, t_1]$.

As a result of the above lemma and the main result of the previous section, we know that we can write

$$V^*[x_1, t_1] = x_1'P^*(t_1)x_1 \tag{3.7}$$

for some $P^*(t_1)$, all x_1 and all $t_1 \in (t_0, t_f]$. Next, we demonstrate an inequality for $P^*(\cdot)$ and the fact that it has bounded variation.

<u>Lemma II.3.2</u>: With the hypotheses of Theorem II.3.1, let $P^*(\cdot)$ be defined by (3.7). Then with $q(x, u) = x'Qx + 2x'Hu + u'Ru$,

$$\int_{t_1}^{t_2} q(x, u)dt + x'(t)P^*(t)x(t) \Big|_{t=t_1}^{t=t_2} \geq 0 \tag{3.8}$$

for all $[t_1, t_2] \subseteq (t_0, t_f]$ and all $x(t_1) = x_1$, and $P^*(\cdot)$ is of bounded variation on $(t_0, t_f]$.

<u>*Proof*</u>: In view of (3.7), (3.8) is nothing but the principle of optimality:

$$V^*[x_1, t_1] \leq \int_{t_1}^{t_2} q(x, u)dt + V^*[x_2, t_2].$$

Next, let $W(t)$ be the solution on $(t_0, t_f]$ of

$$\dot{W} + WF + F'W + Q = 0 \qquad W(t_f) = S, \tag{3.9}$$

let $x(t_1)$ be arbitrary, and let $u(\cdot)$ be zero on $[t_1, t_2]$. It follows that

$$\int_{t_1}^{t_2} q(x, 0)dt = -x'(t)W(t)x(t) \Big|_{t=t_1}^{t=t_2}.$$

...ng that $x(t_2) = \Phi(t_2, t_1)x(t_1)$, we conclude from (3.8) that

$$x'(t_1) \ [\Phi'(t_2, t_1)[P^*(t_2) - W(t_2)]\Phi(t_2, t_1)]x(t_1)$$

$$\geq x'(t_1)[P^*(t_1) - W(t_1)]x(t_1).$$

...e this holds for all $x(t_1)$, it follows easily that $\Phi'(t, t_f)[P^*(t) - W(t)]\Phi(t, t_f)$ ...onotone non decreasing on $(t_0, t_f]$. The same matrix is accordingly of bounded ...ation; it is trivial then that $P^*(\cdot)$ has this property.

To complete the proof of Theorem II.3.1, we need to show that $P^*(t_f) \leq S$ and ...$^*) \geq 0$ within $(t_0, t_f]$. That $P^*(t_f) \leq S$ follows trivially. In the next lemma, ...rove a result close to $dM(P^*) \geq 0$; the result is restricted in that the vector ... of the $dM(P^*) \geq 0$ definition must be related to $u(\cdot)$.

Lemma II.3.3: Let $P(t)$ be any $n \times n$ matrix symmetric and of bounded variation on $[t_1, t_2] \subset [t_0, t_f]$, let $u(\cdot)$ be any piecewise continuous control on $[t_1, t_2]$ and let $x(t_1)$ be the state at time t_1 of (3.1). Then

$$\int_{t_1}^{t_2} [x'(t) \ \ u'(t)] \begin{bmatrix} dP + (PF + F'P)dt & PGdt \\ G'Pdt & 0 \end{bmatrix} \begin{bmatrix} x(t) \\ u(t) \end{bmatrix} = x'(t)P(t)x(t) \ \Big|_{t=t_1}^{t=t_2} \tag{3.10}$$

and with the hypothesis of Theorem II.3.1 and the definition (3.7),

$$\int_{t_1}^{t_2} [x'(t) \ \ u'(t)]dM(P^*) \begin{bmatrix} x(t) \\ u(t) \end{bmatrix} \geq 0 \ . \tag{3.11}$$

...f: The proof of (3.10) relies on a standard "integration by parts" result and the ...etry of $P(t)$. The continuity of $x(\cdot)$, needed for this integration by parts, ...ows from the piecewise continuity of $F(\cdot)$, $G(\cdot)$ and $u(\cdot)$. Equation (3.11) is ...mmediate consequence of (3.8) and (3.10).

The main result of [4] is as stated in the theorem, save that the condition ...$^*) \geq 0$ is replaced by the requirement that (3.11) hold for all $x(t_1)$ and $u(\cdot)$, ... $x(\cdot)$ defined by (3.1). The last lemma in the chain proving Theorem II.3.1 ...s that this restriction is not necessary.

Lemma II.3.4: With notation as defined above, let $P^*(\cdot)$ be such that (3.11) holds for all $x(t_1)$ and $u(\cdot)$. Then $dM(P^*) \geq 0$.

...f: Suppose that $dM(P^*) \geq 0$ fails. Then for some $\varepsilon_1 > 0$, $[t_\alpha, t_\beta] \subset (t_0, t_f]$ ... $u(\cdot)$ and $v(\cdot)$ with appropriate continuity properties, one has

$$\int_{t_\alpha}^{t_\beta} [v'(t) \ \ u'(t)]dM(P^*) \begin{bmatrix} v(t) \\ u(t) \end{bmatrix} < -\varepsilon_1 \ .$$

strategy to deduce a contradiction will be to show that there exists a partition of $[t_\alpha, t_\beta]$ into intervals $[t_\alpha, \tau_2], [\tau_2, \tau_3], \ldots [\tau_{N-1}, t_\beta]$ such that

$$\int_{t_\alpha}^{t_\beta} [v'(t) \quad u'(t)] dM(P^*) \begin{bmatrix} v(t) \\ u(t) \end{bmatrix}$$

can be approximated by

$$\sum_{i=1}^{N-1} \int_{\tau_i}^{\tau_{i+1}} [x_i'(t) \quad u'(t)] dM(P^*) \begin{bmatrix} x_i(t) \\ u(t) \end{bmatrix}$$

where the $x_i(\cdot)$ are all state trajectories of $\dot{x} = Fx + Gu$. Since the approximating quantity must be nonnegative, a contradiction obtains.

The details of the argument follow.

Because $u(\cdot)$ is piecewise continuous, $||\Phi(t, \tau)G(\tau)u(\tau)||$ is bounded on $[t_\alpha, t_\beta] \times [t_\alpha, t_\beta]$, and accordingly, given arbitrary $\varepsilon_2 > 0$, there exists a δ_1 such that

$$\left|\left| \int_{t-\delta_1}^{t} \Phi(t, \tau)G(\tau)u(\tau)d\tau \right|\right| < \tfrac{1}{2} \varepsilon_2$$

for all $[t-\delta_1, t] \subset [t_\alpha, t_\beta]$. Also, because $v(\cdot)$ is continuous on $[t_\alpha, t_\beta]$, there exists a δ_2 such that

$$\sup_{\tau \in [t-\delta_2, t]} ||v(t) - \Phi(t, \tau)v(\tau)|| < \tfrac{1}{2} \varepsilon_2$$

for all $[t-\delta_2, t] \subset [t_\alpha, t_\beta]$. Let $\delta = \min(\delta_1, \delta_2)$. Choose a finite set of points $A_\delta = \{\tau_1 = t_\alpha, \tau_2, \ldots, \tau_N = t_\beta\}$ with $\tau_i < \tau_{i+1}$ such that $\tau_{i+1} - \tau_i < \delta$ and such that P^* does not have a jump at τ_i. (The latter requirement can be fulfilled since $P^*(\cdot)$ as a function of bounded variation is differentiable almost everywhere). Now define

$$x_i(\tau_i) = v(\tau_i)$$

$$x_i(t) = \Phi(t, \tau_i)x_i(\tau_i) + \int_{\tau_i}^{t} \Phi(t, \tau)G(\tau)u(\tau)dt$$

for $\tau_i < t < \tau_{i+1}$. Observe that on $[\tau_i, \tau_{i+1})$, $x_i(\cdot)$ is a state trajectory corresponding to $u(\cdot)$, and also $||x_i(\cdot) - v(\cdot)|| < \varepsilon_2$ by virtue of the definition of δ.

Now we have

$$\int_{\tau_i}^{\tau_{i+1}} [v'(t) \quad u'(t)] dM(P^*) \begin{bmatrix} v(t) \\ u(t) \end{bmatrix} = \int_{\tau_i}^{\tau_{i+1}} [x_i'(t) \quad u'(t)] dM(P^*) \begin{bmatrix} x_i(t) \\ u(t) \end{bmatrix}$$

$$+ 2 \int_{\tau_i}^{\tau_{i+1}} [v'(t)-x_i'(t) \quad 0] dM(P^*) \begin{bmatrix} x_i(t) \\ u(t) \end{bmatrix}$$

$$+ \int_{\tau_i}^{\tau_{i+1}} [v'(t)-x_i'(t) \quad 0] dM(P^*) \begin{bmatrix} v(t)-x_i(t) \\ 0 \end{bmatrix}.$$

magnitude of the second and third terms can be overbounded by a quantity involving total variation of P^* on $[\tau_i, \tau_{i+1}]$ and ϵ_2; moreover, the bound approaches ɔ as ϵ_2 approaches zero. Collecting the $[\tau_i, \tau_{i+1}]$ intervals, we conclude that

$$\left| \int_{t_\alpha}^{t_\beta} [v(t) \quad u'(t)] dM(P^*) \begin{bmatrix} v(t) \\ u(t) \end{bmatrix} - \sum_{i=1}^{N-1} \int_{\tau_i}^{\tau_{i+1}} [x_i'(t) \quad u(t)] dM(P^*) \begin{bmatrix} x_i(t) \\ u(t) \end{bmatrix} \right|$$

$$\leq K\epsilon_2$$

some K reflecting, inter alia, the total variation of P^* on $[t_\alpha, t_\beta]$. Choosing such that $K\epsilon_2 < \epsilon_1$, we obtain a contradiction to the facts that

$$\int_{t_\alpha}^{t_\beta} [v'(t) \quad u'(t)] dM(P^*) \begin{bmatrix} v(t) \\ u(t) \end{bmatrix} < - \epsilon_1$$

$$\int_{\tau_i}^{\tau_{i+1}} [x_i'(t) \quad u(t)] dM(P^*) \begin{bmatrix} x_i(t) \\ u(t) \end{bmatrix} \geq 0$$

all $[\tau_i, \tau_{i+1}] \subset (t_0, t_f)$.
The chain of reasoning proving Theorem II.3.1 is now completed.

rks 3.1: 1. The condition $V[0, t_0, u(\cdot)] \geq 0$ is equivalent to $V^*[0, t_0] = 0$. $V[0, t_0, u(\cdot)] < 0$ for some particular $u(\cdot)$, scaling that $u(\cdot)$ scales V, so V can be made arbitrarily negative, i.e. $V^*[0, t_0] = -\infty$. Hence $V^*[0, t] > -\infty$ ies $V[0, t_0, u(\cdot)] \geq 0$ for all $u(\cdot)$. Next, $V[0, t_0, u(\cdot)] \geq 0$ for all $u(\cdot)$ the observation $V[0, t_0, u(t) \equiv 0] = 0$ shows that $V^*[0, t_0] = 0$).

2. For future reference, we summarize the above result in loose but itively helpful language:

Nonnegativity + controllability $\implies$ V^* finite within
$$(t_0, t_f]. \tag{3.12}$$

Nonnegativity + controllability $\implies \exists P$ such that
$$P(t_f) \leq S \quad \text{and}$$
$$dM(P) \geq 0 \quad \text{within}$$
$$(t_0, t_f]. \tag{3.13}$$

3. The proof of the second part of the theorem proceeded by exhibiting rticular $P(\cdot)$, viz. $P = P^*$ where $V^*[x_1, t_1] = x_1' P^*(t_1) x_1$, satisfying the ed constraints. The reader should be aware however that there are normally other ices P, different from P^*, satisfying the constraints. This point will be ored later in the chapter.

4. It is possible to have a situation in which the nonnegativity and controllability conditions holds, and for which $V^*[x_0, t_0] = -\infty$ for all nonzero x_0. This shows the futility of attempting to improve (3.12) to the extent of obtaining finiteness of $V^*[x_1, t_1]$ for all x_1 and for all $t_1 \in [t_0, t_f]$. The following example is to be found in [4]. The dynamics are $\dot{x} = u$, $t_0 = 0$, $t_f = \pi/2$ and $V[x_0, 0, u(\cdot)] = \int_0^{\pi/2} (-x^2+u^2)dt$.

We first construct a sequence of piecewise continuous controls such that $V[x_0, 0, u_n(\cdot)] \to -\infty$ for any nonzero x_0. Let ε_n be a monotone decreasing sequence with $\varepsilon_n \le 1$ and $\varepsilon_n \to 0$ as $n \to \infty$. Define $u_n(t) = 0$ on $[0, \varepsilon_n)$, $= x_0(\mathrm{cosec}\ \varepsilon_n)$ $(\cos t)$ on $[\varepsilon_n, 1]$. It is easy to verify that $V[x_0, 0, u_n(\cdot)] = -(\varepsilon_n + \cot \varepsilon_n)x_0^2$, which diverges to $-\infty$ as $n \to \infty$.

We next show the nonnegativity of $V[0, 0, u(\cdot)]$. With $u(\cdot)$ piecewise continuous, $x(\cdot)$ is continuous on $[0, \pi/2]$. Define $\bar{x}(\cdot) = [0, \pi]$ by reflecting $x(\cdot)$ in $t = \pi/2$, i.e. $\bar{x}(t) = x(t)$ for $0 \le t \le \pi/2$, $= x(\pi-t)$ for $\pi/2 \le t \le \pi$. There is then a Fourier series expansion of $\bar{x}(\cdot)$ on $[0, \pi]$ with $\bar{x}(t) = \sum_{k=1}^{\infty} a_k \sin kt$. One then computes $V[0, 0, u(\cdot)] = \frac{\pi}{4}\sum_{k=1}^{\infty} a_k^2(k^2-1)$, from which the nonnegativity is evident. Notice incidentally that all controls of the form $\alpha \sin t$ lead to $V[0, 0, u(\cdot)] = 0$.

5. By taking $v(t) \equiv 0$ in the definition of $dM(P^*) \ge 0$, we obtain independent verification of the fact that $R(t) \ge 0$ is necessary for a finite V^*. As noted earlier, nonnegativity of R has long been recognized as necessary for the existence of a finite optimal performance index.

6. The matrix P^* need not be continuous. Consider for example dynamics $\dot{x} = u$, with $V[x_0, 0, u(\cdot)] = -\int_0^1 k(t)x(t)u(t)dt + \frac{1}{2}x^2(1)$, where $k(t) = 0$ on $[0, \frac{1}{2})$, $k(t) = 1$ on $[\frac{1}{2}, 1]$. It is readily verified that $V[x(t), t, u(\cdot)] = \frac{1}{2}x^2(\frac{1}{2})$ for $t \le \frac{1}{2}$ and $V[x(t), t, u(\cdot)] = \frac{1}{2}x^2(t)$ for $t \ge \frac{1}{2}$. This implies that $V^*[x(t), t] = 0$ for $t < \frac{1}{2}$ and $V^*[x(t), t] = \frac{1}{2}x^2(t)$ for $t \ge \frac{1}{2}$. We can however establish that $P^*(\cdot)$ and indeed any $P(\cdot)$ satisfying the Riemann-Stieltjes inequality can have jumps in only one direction:

Lemma II.3.5: Let $P(t)$ be a matrix, symmetric and of bounded variation on $(t_0, t_f]$ and satisfying the conditions $P(t_f) \le S$ and $dM(P) \ge 0$ within $(t_0, t_f]$. Then all jumps in $P(t)$ are nonnegative i.e.,

$$\lim_{\tau \uparrow t} P(\tau) \le P(t) \quad \text{for}\ t \in (t_0, t_f]$$

$$\lim_{\tau \downarrow t} P(\tau) \ge P(t) \quad \text{for}\ t \in (t_0, t_f).$$

Proof: Let $t \in (t_0, t_f]$ be a point of discontinuity of $P(\cdot)$. Let $v(\cdot)$ be an arbitrary constant and $u(\cdot)$ zero in $[t-\delta, t] \subset (t_0, t_f]$. Using the fact that $dM(P) \ge 0$ on $(t_0, t_f]$ and therefore $[t-\delta, t]$, and letting $\delta \to 0$ yields $v'(t)[P(t) - \lim_{\tau \uparrow t} P(\tau)]v(t) \ge 0$. Since $v(t)$ is arbitrary, the first result follows.

second is proved the same way.

We shall later use without explicit comment trivial variations obtained by sing the interval at t_0 or opening it at t_f.

The proof of a result similar to Theorem 3.1 appears in [3] as an extension of totally singular case studied in [7]. The approach in [7] is to regularize the gular problem, replacing it with a nonsingular one obtained by adding the positive ntity $\varepsilon \int_{t_0}^{T} u'u\, dt$ to the cost in the singular problem and allowing ε to approach o. The nonsingular problem is of course much easier to solve, but one naturally to prove things concerning the limit as $\varepsilon \to 0$. (The same idea is also used in). A cleaner derivation, bypassing the need to obtain conditions for the totally gular case prior to the partially singular case, is to be found in [11], and [12] dulo minor changes such as time reversal); an important feature of the proof is the of the Helly convergence theorem for sequences of functions of bounded variation.

Now we turn to the second main result of this section. As a partial converse to orem 3.1, we have the following:

Theorem II.3.2: Suppose that there exists a symmetric $P(\cdot)$ of bounded variation defined on $[t_0, t_f]$, with $P(t_f) \le S$ and with $dM(P) \ge 0$ within $[t_0, t_f]$. Then $V[0, t_0, u(\cdot)] \ge 0$ for every $u(\cdot)$.

of: By the result of Lemma 3.4 we can write

$$\int_{t_1}^{t_2} [x'(t)\ u'(t)]\, dM(P)\begin{bmatrix} x(t) \\ x(t) \end{bmatrix} = x'(t)P(t)x(t)\Big|_{t=t_1}^{t=t_2} + \int_{t_1}^{t_2} q(x, u)\, dt$$

each $[t_1, t_2] \subseteq [t_0, t_f]$. By assumption $dM(P) \ge 0$ and so for the interval , $t_f]$, we obtain

$$x'(t_0)P(t_0)x(t_0) \le x'(t_f)P(t_f)x(t_f) + \int_{t_0}^{t_f} q(x, u)\, dt$$

each $x(t_0)$ and each $u(\cdot)$. In particular, for $x(t_0) = 0$, and noting $P(t_f) \le S$, obtain $V[0, t_0, u(\cdot)] \ge 0$ for each $u(\cdot)$.

arks 3.2: 1. Summarizing the result in loose language, we have

$$P(t_f) \le S$$
and $dM(P) \ge 0 \qquad \implies$ nonnegativity condition $\cdot$ \hfill (3.14)
within $[t_0, t_f]$

2. Statements (3.12) through (3.14) highlight the extent to which orems 3.1 and 3.2 fail to be complete converses. There are basically two aspects this failure, one residing in the need for controllability in Theorem 3.1 and its ence in Theorem 3.2, the other resulting from the fact that a left-closed interval dition is required to guarantee nonnegativity, which only implies a left-open

interval condition.

      3. In fact the hypotheses of Theorem 3.2 imply the slightly stronger result that $V^*[x_0, t_0]$ is finite for all x_0. This follows easily on working with the inequality contained in the proof of the theorem.

      In order to get tidier results (with tidiness measured by the occurrence of conditions which are both necessary and sufficient, not one or the other), we suggest a change of viewpoint, based on the observations of Section 1 concerning robust and non-robust problems. Thus we should be interested in not merely conditions for $V^*[0, t_0]$ to be finite, but in conditions for $V^*[0, t^\prime]$ to be finite for $t^\prime$ in a neighborhood of t_0 (robustness as far as initial time is concerned), and in conditions for $V^*[x_0, t_0]$ to be finite for x_0 in a neighborhood of 0. (Because of the linear-quadratic nature of the problem, this means that $V^*[x_0, t_0]$ is finite for all x_0).

      One step in this direction is provided by Theorem 3.3 below, which connects finiteness of $V^*[x_0, t_0]$ for all x_0 with the existence of a matrix P satisfying certain conditions.

      <u>Theorem II.3.3</u>: $V^*[x_0, t_0] > -\infty$ for all x_0 if and only if there exists on $[t_0, t_f]$ a symmetric $P(\cdot)$ of bounded variation such that $P(t_f) \leq S$ and $dM(P) \geq 0$ within $[t_0, t_f]$.

Proof: Since $V[0, t_0, u(\cdot)] \geq 0$ for all $u(\cdot)$ if and only if $V^*[0, t_0] = 0$, we know from Theorem 3.1 that $V^*[x_1, t_1] > -\infty$ for all x_1 and $t_1 \in (t_0, t_f]$. Since also $V^*[x_0, t_0] > -\infty$ for all x_0, we have $V^*[x_1, t_1] = x_1^{\hat{}}P^*(t_1)x_1$ for some symmetric $P^*(t_1)$, all x_1 and all $t_1 \in [t_0, t_f]$. Trivial variation on the lemmas used in proving Theorem 3.1 yields the necessity claim of the theorem. Sufficiency is a simple consequence of the proof of Theorem 3.2, as noted in Remark 2.3.3.

      The first necessity and sufficiency result on the existence of $V^*[x_0, t_0]$ appears to be that of [6]. It is the same as that of Theorem 3.3, save that the condition $dM(P) \geq 0$ is replaced by the restricted condition (3.11).

      <u>Remarks 3.3</u>: 1. It is an immediate consequence of this result that if $V^*[x_0, t_0] > -\infty$ for all x_0, then $V^*[x_1, t_1] > -\infty$ for all x_1 and all $t_1 \in [t_0, t_f]$. (Simply use the fact that if $dM(P) \geq 0$ on all closed intervals contained in $[t_0, t_f]$, then $dM(P) \geq 0$ on all closed intervals contained in $[t_1, t_f]$).

      2. We describe this result as extended by Remark 1, as

$$V^* \text{ finite on } [t_0, t_f] \iff P(t_f) \leq S$$

$$\text{and} \quad dM(P) \geq 0 \quad \text{within} \quad [t_0, t_f]$$

$$\iff V^* \text{ finite within } [t_0, t_f]. \qquad (3.15)$$

      3. A comparison of Theorems 3.1 and 3.3 shows that the interval within which $dM(P) \geq 0$ is open or closed at t_0 according as we restrict the values of $x(t_0)$ of interest, or $x(t_0)$ is free. It will be seen subsequently that the interval

en or closed at t_f according as $x(t_f)$ is or is not restricted.

4. The example of Remarks 3.1 shows, in conjunction with Theorem 3.3,
it is possible to have a symmetric P of bounded variation, satisfying $P(t_f) \leq S$
with $dM(P) \geq 0$ within $(t_0, t_f]$ but not within $[t_0, t_f]$.
We now state two corollaries to Theorem 3.3 which show that the necessary and
-cient conditions of that theorem are indeed just generalizations of better known,
-ess general, conditions for the existence of a solution to the control problem.
he first of these corollaries, which is virtually self-evident, we assume that
P matrix is differentiable on some $[t_1, t_2] \subseteq [t_0, t_f]$ and obtain a linear
-x differential inequality. For the second we assume that the problem is nonsing-
on the interval $[t_0, t_f]$; it is then possible to show that the necessary and
-cient conditions of Theorem 3.3 are equivalent to the well-known condition that
satisfy the Riccati differential equation on $[t_0, t_f]$. The interested reader
consult [4], [5] and [10] for these corollaries and other closely related results.

-lary II.3.1: Let $P(t)$ be a matrix, symmetric and of bounded variation on
$t_f]$ with $dM(P) \geq 0$ within $[t_0, t_f]$. Further suppose that $P(t)$ is different-
- in a neighborhood of t. Then

$$\begin{bmatrix} \dot{P} + PF + F'P + Q & PG + H \\ (PG+H)' & R \end{bmatrix} \geq 0$$

neighborhood of t. Conversely, satisfaction of this inequality in a neighborhood
: implies that $dM(P) \geq 0$ within this neighborhood.

-lary II.3.2: Assume that $R(t) > 0$ on $[t_0, t_f]$. If $V^*[x_0, t_0]$ is finite for
x_0, then the matrix $P^*(t)$ defined by $V^*[x(t), t] = x'(t)P^*(t)x(t)$ on $[t_0, t_f]$
ntinuously differentiable and satisfies

$$\dot{P}^* + P^*F + F'P^* + Q - (P^*G+H)R^{-1}(P^*G+H)' = 0, \quad P^*(t_f) = S$$

$[t_0, t_f]$. Conversely, if the solution of this Riccati equation has no escape time
$[t_0, t_f]$, $V^*[x_0, t_0]$ is finite for each x_0 and is given by $x_0'P^*(t_0)x_0$.

A variant of Corollary II.3.1 has found extensive use in problems of time-varying
ork synthesis and covariance factorization, see [11, 12].
This completes our discussion of the idea of "robustness with respect to initial
-". We turn now to a consideration of "robustness with respect to initial time",
the goal of connecting the notions of initial state and time robustness. Remark
- considered a change of initial time from t_0 to some $t \in (t_0, t_f]$. We need
:o consider the possibility of taking an initial time $t_{-1} < t_0$. The next theorem
-des the main result.

Theorem II.3.4: Suppose that $V^*[x_0, t_0] > -\infty$ for all x_0. Then there exists

$t_{-1} < t_0$ and definitions on $[t_{-1}, t_0)$ of $F(\cdot)$, $G(\cdot)$, $H(\cdot)$, $Q(\cdot)$ and $R(\cdot)$ such that these quantities are continuous on $[t_{-1}, t_f]$, such that

$$\int_{t_{-1}}^{t_1} \Phi(t_1, \tau) G(\tau) G'(\tau) \Phi'(t_1, \tau) d\tau > 0 \qquad \forall \; t_1 \in (t_{-1}, t_f], \qquad (3.16)$$

such that $V[0, t_{-1}, u(\cdot)] \geq 0$ for all $u(\cdot)$, and in fact $V^*[x_{-1}, t_{-1}] > -\infty$ for all $x(t_{-1}) = x_{-1}$.

Proof: We consider the proof of the theorem first for a special situation; then we show that the general situation can always be reduced to the special situation.

Let $P(\cdot)$ be the matrix whose existence on $[t_0, t_f]$ is guaranteed by Theorem 3.3, and suppose for the moment that $\dot{P}$ exists in a neighborhood $[t_0, t_0+\varepsilon]$ of t_0. Take $t_{-1} < t_0$ and otherwise arbitrary, and take $F(\cdot)$, $G(\cdot)$ on $[t_{-1}, \frac{1}{2}(t_{-1} + t_0)]$ to be any constant, completely controllable pair; define $F(\cdot)$, $G(\cdot)$ on $(\frac{1}{2}(t_{-1} + t_0), t_0)$ so as to ensure smooth joins, and continuity on $[t_{-1}, t_f]$.

Let $\dot{P}(t) = \dot{P}(t_0)$ on $[t_{-1}, t_0)$; this ensures that $P(t)$ is continuous on $[t_{-1}, t_0]$. Choose $Q(\cdot)$ on $[t_{-1}, t_0)$ such that $\dot{P} + PF + F'P + Q$ is constant on $[t_{-1}, t_0]$; this ensures $Q(\cdot)$ is continuous on $[t_{-1}, t_f]$. Choose $H(\cdot)$ on $[t_{-1}, t_0)$ so that $PG + H$ is constant on $[t_{-1}, t_0]$; again, $H(\cdot)$ is continuous on $[t_{-1}, t_f]$. Finally, choose $R(\cdot)$ on $[t_{-1}, t_0)$ to be constant and equal to $R(t_0)$, ensuring thereby continuity on $[t_{-1}, t_f]$.

These choices guarantee that

$$\dot{M} = \begin{bmatrix} \dot{P} + PG + F'P + Q & PG + H \\ (PG + H)' & R \end{bmatrix}$$

is constant on $[t_{-1}, t_0]$. The fact that $dM(P) \geq 0$ on $[t_0, t_1]$ for all $t_1 \in [t_0, t_f]$ and that $\dot{P}$ exists in a neighborhood of t_0 ensures that $\dot{M}(t_0) \geq 0$ — see Corollary 3.1. Consequently $\dot{M} \geq 0$ on $[t_{-1}, t_0]$ and then $dM(P) \geq 0$ within $[t_{-1}, t_f]$. By Theorem 3.3, $V^*[x_{-1}, t_{-1}] > -\infty$ for all x_{-1}, while (3.16) holds because of the choice of $F(\cdot)$, $G(\cdot)$ on $[t_{-1}, \frac{1}{2}(t_{-1} + t_0)]$.

Now suppose that $\dot{P}$ does not exist in a neighborhood of $[t_0, t_0+t]$ of t_0. Consider the following equation for $t \leq t_0$:

$$\dot{P} + PF + F'P + Q - (PG + H)[R_0 + (t_0 - t)^{\frac{1}{2}}I]^{-1}(PG + H)' = 0 \qquad (3.17)$$

where $R_0 = R(t_0)$, and $F(\cdot)$, $G(\cdot)$, $H(\cdot)$ and $Q(\cdot)$ are arbitrary continuous extensions of these quantities into $t < t_0$, chosen to ensure continuity at t_0. Equation (3.17) is initialised by the known quantity $\Pi_0 = P(t_0)$.

In case R_0 is nonsingular, $P(t)$ is guaranteed to exist in some interval $[t_{-2}, t_0]$, with $\dot{P}$ guaranteed to exist in $[t_{-2}, t_0)$. However, in case R_0 is singular, $\dot{P}$ is unbounded as $t \uparrow t_0$, and so an existence question arises, which we now resolve. Set $\tau = (t_0 - t)^{\frac{1}{2}}$. Then

$$\frac{dP}{dt} = \frac{dP}{d\tau}\frac{d\tau}{dt} = -\frac{1}{2(t_0 - t)^{\frac{1}{2}}}\frac{dP}{d\tau}$$

with τ the new independent variable, (3.17) becomes

$$-\frac{dP}{d\tau} + 2\tau[PF + F'P + Q] - (PG + H)2\tau(R_0 + \tau I)^{-1}(PG + H)' = 0. \qquad (3.18)$$

equation is defined in the interval $\tau \geq 0$; strictly, we should have used diff-

t symbols for $F(\cdot)$, etc., to reflect their change of independent variable. The

tion has $P(\tau)|_{\tau=0} = \Pi_0$. Now $\tau(R_0 + \tau I)^{-1}$ is obviously continuous for $\tau > 0$,

it is not hard to check that it is continuous at $\tau = 0$. Therefore, $P(\tau)$ exists

ome interval $[0, \tau_2]$ with $\frac{dP(\tau)}{d\tau}$ continuous there. It follows that (3.17) has

lution in some interval $[t_{-2}, t_0]$ with $\frac{dP(t)}{dt}$ existing on $[t_{-2}, t_0)$ and, in

, in a neighborhood of t_{-2}.

Now (3.17) implies that

$$\begin{bmatrix} \dot{P} + PF + F'P + Q & PG + H \\ (PG + H)' & R \end{bmatrix} \geq 0$$

$[t_{-2}, t_0)$, where $R = R_0 + (t_0 - t)^{\frac{1}{2}}I$ and is nonsingular. Since $\lim\limits_{t\uparrow t_0} P(t) = \Pi_0$

$t_0)$, it is clear that $dM(P) \geq 0$ within $[t_{-2}, t_f]$. Now we can use the first

of the proof to further extend on $[t_{-1}, t_{-2}]$, since $\dot{P}(t)$ exists in a neighbor-

of t_{-2}. In this way, the controllability assumption is fulfilled, and the

rem is proved.

rks 3.4: 1. This theorem is the first in the book to introduce an <u>extendability</u>

erion. The first use of the extendability idea of which we are aware is in [1],

e nonsingular problems only were discussed.

    2. In case $R(t)$ is nonsingular throughout $[t_0, t_f]$, the above theorem

uch easier to prove, for $V^*[x_1, t_1] = x_1'P^*(t_1)x_1$ with $P^*(t_1)$ the solution of a

ati equation, $t_1 \in [t_0, t_f]$. Then $\dot{P}(t_1)$ automatically exists for all

$[t_0, t_f]$.

    3. We summarise the result as:

V^* finite on $[t_0, t_f] \Longrightarrow$ nonnegativity and controllability

                        for the extended interval

                        $[t_{-1}, t_f]$ (3.19)

V^* finite on $[t_0, t_f] \Longrightarrow V^*$ finite on extended interval

                        $[t_{-1}, t_f]$.   (3.20)

    4. An examination of the proof of Theorem 3.4 will show that t_{-1} may

aken arbitrarily close to t_0. <u>This fact essentially makes (3.19) and (3.12) con-</u>

verse statements; the converse to (3.13) is obtained by replacing V^* finite on $[t_0, t_f]$ in (3.19) by the equivalent statement involving P contained in (3.15).

5. The result contained in the preceding theorem might lead one to make the following conjecture, which we can readily show is false: suppose that $V^*[x_0, t_0] > -\infty$ for all x_0, and that $F(\cdot)$, $G(\cdot)$, $H(\cdot)$, $Q(\cdot)$, $R(\cdot)$ are defined on $[t_{-1}, t_0]$ such that these quantities are continuous on $[t_{-1}, t_f]$; then there exists $t_{-2} \in [t_{-1}, t_0)$ such that $V^*[x_{-2}, t_2] > -\infty$. (Effectively, it is being claimed that the set of t_0 for which $V^*[x_0, t_0] > -\infty$ is open). By way of counterexample, consider $\dot{x} = u$, $V[x(\tau), \tau, u(\cdot)] = \int_{\tau}^{t_f} [xu + \rho(t)u^2] dt$ where $\rho(t) = 0$, $t \in [t_0, t_f]$ and $\rho(t) < 0$ for $t \in [t_{-1}, t_0)$, with $\rho(\cdot)$ continuous on $[t_{-1}, t_f]$. Certainly then, it is impossible that $V^*[x_{-2}, t_{-2}] > -\infty$ for $t_2 \in [t_{-1}, t_0)$, since ρ is negative on $[t_{-2}, t_0)$. However, $V[x(t_0), t_0, u(\cdot)] = \frac{1}{2} x^2(t_f) - \frac{1}{2} x^2(t_0)$ for all $u(\cdot)$, so that $V^*[x_0, t_0] = -\frac{1}{2} x_0^2$.

A second example, in which negative $\rho(t)$ is not used is provided by $\dot{x} = g(t)u$, $V[x(\tau), \tau, u(\cdot)] = \int_{\tau}^{t_f} xu \, dt$ where $g(t) = 0$ for $t < 0$ and $g(t) = 1$ for $t \geq 0$. It is unclear whether such examples can be constructed in case $R(t) \geq 0$ and F, G, H, Q, R are all continuous.

Under the restriction $R(t) > 0$ on $[t_0, t_f]$ and a continuity requirement on $R(\cdot)$, the above conjecture is definitely true, for $V^*[x(\tau), \tau]$ is defined via the solution to a Riccati equation which, if it exists at t_0, exists in a neighborhood around t_0, including points to the left of t_0.

In this section we have so far separately considered robustness with respect to the initial state (Theorem 3.3) and robustness with respect to the initial time (Theorem 3.4). It is now convenient to summarize these theorems together with Theorem 3.1 as

Theorem II.3.5: With notation as previously, the following conditions are equivalent:

(a) $V^*[x_0, t_0]$ is finite for all x_0.

(b) $V^*[x(t), t]$ is finite for all $x(t)$ and for all $t \in [t_0, t_f]$.

(c) There exist extensions of the interval of definition of $F(\cdot)$, etc., such. that $V[0, t_{-1}, u(\cdot)] \geq 0$ for some $t_{-1} < t_0$, and with controllability on $[t_{-1}, t]$ for all $t \in (t_{-1}, t_f]$.

(d) There exists on $[t_0, t_f]$ a symmetric $P(\cdot)$ of bounded variation with $P(t_f) \leq S$ and $dM(P) \geq 0$ within $[t_0, t_f]$.

We emphasise the fact that the conditions involving the Riemann-Stieltjes integral are simultaneously necessary and sufficient.

We conclude this section with remarks of minor significance on another type of perturbation. To this point, we have considered the effect of perturbations of the initial state away from zero, and perturbations of the initial time t_0. Another type of perturbation that can be considered is a perturbation of the underlying matrices

H, Q and R. If R(t) is nonsingular on $[t_0, t_f]$, existence of a solution

certain Riccati equation on $[t_0, t_f]$ is necessary and sufficient for

, $t_0] > -\infty$ for all x_0, and this existence condition is robust with respect to

tions in F, G, H, Q and R which are suitably small. If the Riccati equation

solution on $(t_0, t_f]$ with escape time at t_0, so that $V^*[x_0, t_0] = 0$ for

0 but is $-\infty$ for some $x_0 \neq 0$, matters are not quite the same; variation in

tc. may cause $V^*[x_0, t_0] > -\infty$ for all x_0, or $V^*[x_1, t_1] = -\infty$ for some x_1

$t_1 > t_0$ with t_1 close to t_0. A third possibility arises if $R(t_1)$ is sing-

for some $t_1 \in [t_0, t_f)$. Then perturbations can make $R(t_1)$ indefinite, and

inly then $V^*[x_1, t_1]$ may be $-\infty$; this will be the case even if $V^*[x_0, t_0] > -\infty$

ll x_0 prior to perturbation.

Evidently, two crucial issues affecting tolerance of perturbations are whether

is nonsingular on $[t_0, t_f]$, or singular somewhere in the interval, and whether

, $t_0] > -\infty$ for all x_0, or $V[0, t, u(\cdot)] \geq 0$ for all $u(\cdot)$, with $V[x_0, t_0]$

failing for some x_0. In this latter case, we can establish a result of minor

quence which applies both to nonsingular and singular $R(\cdot)$ cases; it states

a perturbation in $R(\cdot)$ can always be found to ensure $V^*[x_0, t_0] > -\infty$ for all

Theorem II.3.6: Suppose that $V[0, t_0, u(\cdot)] \geq 0$ for all $u(\cdot)$ and that the

controllability condition (3.5) holds. Suppose that $V^*[x_0, t_0] > -\infty$ fails for

some x_0. Let $t_1 \in (t_0, t_f]$ be arbitrary (in particular, t_1 may be arbit-

rarily close to t_0), and let $\varepsilon > 0$ be arbitrary. Let $\rho(t)$ be continuous on

$[t_0, t_f]$ with $\rho(t_0) = 1$, $\rho(t) > 0$ on $[t_0, t_1)$, $\rho(t) = 0$ on $[t_1, t_f]$. With

$R(t)$ replaced by $\bar{R}(t) = R(t) + \varepsilon \rho(t)I$ on $[t_0, t_f]$, $\bar{V}^*[x_0, t_0] > -\infty$ for all

x_0.

Let $t_2 \in (t_0, t_1)$. With $x(t_0) = 0$, we have

$$0 \leq \int_{t_0}^{t_2} \{x'Q(t)x + 2u'H(t)x + u'R(t)u\}dt$$
$$+ \int_{t_2}^{t_f} x'Q(t)x + 2u'H(t)x + u'R(t)u\}dt$$
$$+ x'(t_f Sx(t_f)$$

)

$$0 \leq \int_{t_0}^{t_2} \{x'Q(t)x + 2u'H(t)x + u'R(t)u\}dt$$
$$+ x'(t_2)P^*(t_2)x(t_2)$$

$x'(t_2)P^*(t_2)x(t_2) = V^*[x(t_2), t_2]$.

A straightforward argument, set out in the main result of [1], show that the prob-

minimizing

$$\int_{t_0}^{t_2}\{x'Q(t)x + 2u'H(t)x + u'R(t)u + \varepsilon\rho(t)u'u\}dt$$

$$+ x'(t_2)P^*(t_2)x(t_2)$$

with arbitrary $x(t_0) = x_0$, has for all x_0 a solution which is not $-\infty$, since there exists $\eta > 0$ such that with $x(t_0) = 0$, one has for all $u(\cdot)$

$$\int_{t_0}^{t_2}\{x'Q(t)x + 2u'H(t)x + u'R(t)u + (\varepsilon\rho(t)-\eta)u'u\}dt$$

$$+ x'(t_2)P^*(t_2)x(t_2) \geq 0.$$

Hence for all x_0,

$$\bar{V}^*[x_0, t_0] = \min_{u(\cdot)} \int_{t_0}^{t_2}\{x'Q(t)x + 2u'H(t)x + u'R(t)u$$

$$+ \varepsilon\rho(t)u'u\}dt + \bar{V}^*[x(t_2), t_2]$$

$$\geq \min_{u(\cdot)} \int_{t_0}^{t_2}\{x'Q(t)x + 2u'H(t)x + u'R(t)u$$

$$+ \varepsilon\rho(t)u'u\}dt + V^*[x(t_2), t_2)]$$

$$> -\infty.$$

(The first equality follows from the principal of optimality and the first inequality by monotonicity of $\bar{V}^*$ with ε).

4. ROBUSTNESS IN PROBLEMS WITH END-POINT CONSTRAINTS

Throughout this section, we study the system (3.1) with performance index (3.2). As earlier, $x(t_0)$ is fixed but arbitrary; now $x(t_f)$ is no longer free but constrained by

$$E_f x(t_f) = 0 \tag{4.1}$$

where E_f is a matrix with full row rank, sometimes specialised to the identity. Of course, we are interested in minimizing (3.2), or (3.2) with t_0 replaced by variable t.

We shall begin by reviewing known results drawn from [2-4]. These results suffer from a degree of asymmetry - necessary conditions are not quite sufficient conditions. Then we shall observe that by introducing robustness requirements, this asymmetry can be removed. A new form of robustness enters the picture, additional rather than alternative to those encountered earlier.

The following result is drawn from [2-4]. Its proof can be obtained similarly to the proofs of Theorems 2.1 through 2.3.

Theorem II.4.1: Assume that

$$E_f \int_t^{t_f} \Phi(t_f, \tau) G(\tau) G'(\tau) \Phi'(t_f, \tau) d\tau \, E_f' > 0 \tag{4.2}$$

for all $t \in [t_0, t_f)$. Let Z be a matrix with columns constituting a basis for the nullspace of E_f. A necessary condition for $V[x(t), t, u(\cdot)]$ to have a finite infimum for all $x(t)$ and $t \in [t_0, t_f)$ is that there exists on $[t_0, t_f)$ a symmetric $P(\cdot)$ of bounded variation such that $dM(P) \geq 0$ within $[t_0, t_f)$ and

$$\lim_{t \uparrow t_f} Z'[\Phi'(t, t_f) P(t) \Phi(t, t_f) - S] Z \leq 0. \tag{4.3}$$

A sufficient condition is that a symmetric $P(\cdot)$ of bounded variation exist on $[t_0, t_f]$ with $dM(P) \geq 0$ within $[t_0, t_f]$ and

$$Z'[P(t_f) - S] Z \leq 0. \tag{4.4}$$

arks 5.1: 1. Strictly, [2–4] are concerned with conditions which ensure
(t), t, u(·)] has a finite infimum for all $t \in (t_0, t_f)$ and $V[0, t_0, u(\cdot)] \geq 0$
all u(·). The methods of Section 2 however allow the derivation of Theorem 4.1
the same way that Theorem 3.3 is derived from Theorems 3.1 and 3.2.

2. One $P(\cdot)$ satisfying the necessary conditions is defined by
$)P(t)x(t) = \inf_{u(\cdot)} V[x(t), t, u(\cdot)]$ with $E_f x(t_f) = 0$.

3. The controllability condition (4.2) is the appropriate specialization
a general condition given in Section 2 which ensures that the optimal performance
ex exists, i.e. the state constraint is attainable.

There are at least three distinct ways in which robustness might be sought. First,
can study the effect of allowing variations in t_f, much as we varied the initial
 t_0 in Section 3. Second, we can study the effect of replacing (4.1) by a con-
.on like $||E_f x(t_f)|| \leq \epsilon$ for suitably small ϵ. Third, making use of the idea of
rporating penalty functions [13] in the performance criteria, we can study the
ct of eliminating (4.1) while adding the quantity $Nx'(t_f) E_f' E_f x(t_f)$ to the per-
ance criterion (2.2), with N a large number. Robustness in this latter case
esponds to their existing an optimum performance index for all sufficiently large

As noted in the last section, at least for t_0 replacing t_f, there are problems
ing the first two kinds of robustness. [Strictly, in the last section, the second
 of robustness was viewed as replacing $x(t_0) = 0$ by arbitrary $x(t_0)$, rather
 $||E_0 x(t_0)|| \leq \epsilon$]. Let us now observe that there are also problems lacking the
d kind of robustness. Then we shall go on to discuss the equivalence of the types
obustness. Some of the results are drawn from [14].

Consider $\dot{x} = (t-1)u$, $V[x_0, 0, u(\cdot)] = \int_0^1 xu \, dt$, with side constraint $x(1) = 0$.
how first that, with this constraint, $V \geq 0$ for all $x(0)$ and all piecewise

continuous $u(\cdot)$. Since the constraint is evidently attainable, it follows that inf V exists. Observe that

$$V = \lim_{T\uparrow 1} \int_0^T xu\,dt$$

$$= \lim_{T\uparrow 1} \int_0^T \left\{ -\int_t^1 (\tau-1)u(\tau)d\tau \right\} u(t)\,dt \qquad\qquad [\text{using}\quad x(1) = 0]$$

$$= \lim_{T\uparrow 1} \int_0^T \left\{ \int_t^1 (\tau-1)u(\tau)d\tau \right\} (t-1)^{-1} \frac{d}{dt} \left\{ \int_t^1 (\tau-1)u(\tau)d\tau \right\} dt$$

$$= \lim_{T\uparrow 1} \left\{ \left[(t-1)^{-1} \left[\int_T^1 (\tau-1)u(\tau)d\tau \right]^2 \right]_0^T + \int_0^T u(t) \left[\int_t^1 (\tau-1)u(\tau)d\tau \right] dt \right.$$

$$\left. + \int_0^T (t-1)^{-2} \left[\int_t^1 (\tau-1)u(\tau)d\tau \right]^2 dt \right\} \quad .$$

The second term on the right side is $-V$, whence

$$V = \tfrac{1}{2} \left[\int_0^1 (\tau-1)u(\tau)d\tau \right]^2 + \tfrac{1}{2} \lim_{T\uparrow 1}(T-1)^{-1} \left[\int_T^1 (\tau-1)u(\tau)d\tau \right]^2$$

$$+ \tfrac{1}{2} \lim_{T\uparrow 1} \int_0^T (t-1)^{-2} \left[\int_t^1 (\tau-1)u(\tau)d\tau \right]^2 dt.$$

Using L'Hôpital's rule, the first limit is seen to be zero, the second to be nonnegative. Therefore $V \geq 0$ as claimed.

Now consider the minimization of $V[x_0, 0, u(\cdot); N]$ with the constraint $x(1) = 0$ removed. We shall show there is no finite N for which inf V $> -\infty$. If there were, by Theorem 3.3 there would exist $P(\cdot)$ of bounded variation on $[0, 1]$ with $dM(P) \geq 0$, $P(1) = N$. Such a $P(\cdot)$, if it has jumps, must only have positive jumps as explained in Section 3. Now arguments as in, for example [7], show that for any problem for which $R(t) = 0 \;\forall\, t \in [t_0, t_f]$ and for which $G(\cdot)$ and $H(\cdot)$ are continuous, one has $PG + H = 0 \;\forall\, t \in (t_0, t_f)$ for any P such that $dM(P) \geq 0$ within $[t_0, t_f]$. (The result is easy to establish). This means that here, $P(t)(t-1) + \tfrac{1}{2} = 0$, or $P(t) = \tfrac{1}{2}(1-t)^{-1}$. Satisfaction of the endpoint constraint is accordingly impossible, since $P(\cdot)$ would have an infinitely negative jump there. In this way, we have a contradiction to the claim that inf V $> -\infty$ for some N.

We now turn to the main task of this section, which is to illustrate the equivalence of the three kinds of robustness - robustness with respect to terminal time, terminal state constraint, and terminal weighting matrix in the performance index. The first theorem below shows the equivalence of the last two forms of robustness.

We shall make notational remarks. We recall that

$$v^*[x_0, t_0] = \inf_{u(\cdot)} V[x_0, t_0, u(\cdot)]$$

$$v^*[x_0, \eta_f] = \inf_{u(\cdot)} V[x_0, t_0, u(\cdot)] \qquad \text{subject to} \quad E_f x(t_f) = \eta_f$$

and we define also

$$V[x_0, t_0, u(\cdot); \ N] = V[x_0, t_0, u(\cdot)] + N||E_f x(t_f)||^2$$

$$V^*[x_0, t_0; \ N] = \inf_{u(\cdot)} V[x_0, t_0, u(\cdot); \ N]$$

and

$$V^*_\varepsilon[x_0, t_0] = \inf_{u(\cdot)} V[x_0, t_0, u(\cdot)] \quad \text{subject to} \quad ||E_f x(t_f)|| \leq \varepsilon.$$

Throughout the following t_0 and t_f are considered fixed. Later in the section, we shall consider t_f variable; permitting t_0 variable adds nothing.

Theorem II.4.2: The following conditions are equivalent.
(a) The controllability condition (4.2) holds with t replaced by t_0, and $V^*[x_0, t_0; \ N]$ exists for some N and all x_0.
(b) $V^*[x_0, t_0; \ \bar{N}]$ exists for all $\bar{N} \geq$ some N and is bounded above uniformly in $\bar{N} \geq N$.
(c) $V^*_\varepsilon[x_0, t_0]$ exists for all x_0 and all $\varepsilon > 0$.
(d) $V^*[x_0, \eta_f]$ exists for all x_0 and all η_f.

Moreover, should any one condition hold, we have

$$\lim_{N \to \infty} V^*[x_0, t_0; \ N] = \lim_{\varepsilon \to 0} V^*_\varepsilon[x_0, t_0] = V^*[x_0, \eta_f = 0]. \tag{4.5}$$

Proof: (a) $\Rightarrow$ (b). By the controllability condition, there exists a $u(\cdot)$ taking $x(t_0) = x_0$ to $x(t_f)$ with $E_f x(t_f) = 0$. Then $V[x_0, t_0, u(\cdot)] \geq V^*[x_0, t_0; \ \bar{N}] \geq V^*[x_0, t_0, N]$.

(b) $\Rightarrow$ (c). We show first that $V^*_\varepsilon[x_0, t_0] < \infty$. Suppose this is not the case. The only way this can happen is that if, for some ε, $||E_f x(t_f)|| \leq \varepsilon$ is not attainable. In order to show a contradiction, assume this is the case, and let M be such that $V^*[x_0, t_0; \ \bar{N}] < M - \delta$ for some arbitrary $\delta > 0$ and all $\bar{N}$. Assumption (b) of the Theorem statement guarantees existence of M and δ. Define $U_{\bar{N}}$ as the set of piecewise continuous $u(\cdot)$ for which $V[x_0, t_0, u(\cdot); \ \bar{N}] < M$. Then for $u(\cdot) \in U_{\bar{N}}$ we have

$$V[x_0, t_0, u(\cdot)] < M - \bar{N}||E_f x(t_f)||^2$$

$$< M - \bar{N}\varepsilon^2. \tag{4.6}$$

[We can never have $||E_f x(t_f)|| < \varepsilon$]. Set

$$Q^*_{\bar{N}} = \inf_{u(\cdot) \in U_{\bar{N}}} V[x_0, t_0, u(\cdot)].$$

Since for $\bar{N}_1 > \bar{N}_2$, $U_{\bar{N}_1} \subset U_{\bar{N}_2}$, we see that $Q^*_{\bar{N}}$ is monotone increasing. On the other hand from (4.6) we have

$$Q_{\frac{*}{\bar{N}}} < M - \bar{N} \varepsilon^2$$

from which it is clear that $\lim\limits_{\bar{N}\to\infty} Q_{\frac{*}{\bar{N}}} = -\infty$. Therefore $Q_{\frac{*}{\bar{N}}} = -\infty$ for all $\bar{N}$, and given arbitrary $K > 0$, there exists $\bar{u}(\cdot) \in \mathcal{U}_{\bar{N}+1}$ such that $V[x_0, t_0, \bar{u}(\cdot)] < -K$. Then we have

$$V^*[x_0, t_0; \bar{N}] \leq V[x_0, t_0, \bar{u}(\cdot)] + \bar{N}||E_f\bar{x}(t_f)||^2$$

$$< -K + \bar{N}||E_f\bar{x}(t_f)||^2$$

and

$$V[x_0, t_0, \bar{u}(\cdot); \bar{N}+1] = V[x_0, t_0, \bar{u}(\cdot); \bar{N}] + ||E_f\bar{x}(t_f)||^2$$

$$> V^*[x_0, t_0; \bar{N}] + \frac{1}{N} V^*[x_0, t_0; \bar{N}] + K.$$

Since K is arbitrary, this violates the constraint that $V[x_0, t_0, \bar{u}(\cdot); \bar{N}+1] < M$, and the contradiction is established.

To show that $V_\varepsilon^*[x_0, t_0] > -\infty$ is much easier. We have, for all $u(\cdot)$ such that $||E_f x(t_f)|| \leq \varepsilon$,

$$V^*[x_0, t_0; N] \leq V[x_0, t_0, u(\cdot)] + N \varepsilon^2$$

or

$$V[x_0, t_0, u(\cdot)] \geq V^*[x_0, t_0; N] - N \varepsilon^2.$$

The lower bound on $V_\varepsilon^*[x_0, t_0]$ is immediate.

(c) $\Rightarrow$ (d). First observe that the finiteness of $V_\varepsilon^*[x_0, t_0]$ implies the controllability condition (4.2) holds with t replaced by t_0. For if it did not, there would exist an initial state $\bar{x}_0$ and some value of ε such that for no control $u(\cdot)$ could one ensure that $||E_f x(t_f)|| < \varepsilon$. Now the fact that the controllability condition holds implies that $V^*[x_0, \eta_f] < \infty$, since all η_f are reachable. For fixed η_f, choose ε such that $||\eta_f|| \leq \varepsilon$. Then it is clear that $V^*[x_0, \eta_f] \geq V_\varepsilon^*[x_0, t_0] > -\infty$.

(d) $\Rightarrow$ (a). Controllability is trivial. Next, by Theorem 2.1, we have the representation

$$V^*[x_0, \eta_f] = [x_0' \quad \eta_f'] \begin{bmatrix} P_{00} & P_{0f} \\ P_{f0} & P_{ff} \end{bmatrix} \begin{bmatrix} x_0 \\ \eta_f \end{bmatrix} \qquad (4.7)$$

for some P_{00}, $P_{0f} = P_{f0}'$ and P_{ff}, so that

$$V^*[x_0, t_0; N] = \inf_{\eta_f} \left[\inf_{u(\cdot)} \{V[x_0, t_0, u(\cdot)] + N||\eta_f||^2\} \right]$$

e the class of $u(\cdot)$ are those leading to $E_f x(t_f) = \eta_f$. The inner infimum is
cisely $V^*[x_0, \eta_f] + N||\eta_f||^2$, so that

$$V^*[x_0, t_0; N] = \inf_{\eta_f} \left\{ [x_0' \quad \eta_f'] \begin{bmatrix} P_{00} & P_{0f} \\ P_{f0} & P_{ff}+NI \end{bmatrix} \begin{bmatrix} x_0 \\ \eta_f \end{bmatrix} \right\}$$

$$= x_0'[P_{00} - P_{0f}(P_{ff}+NI)^{-1}P_{f0}]x_0 . \tag{4.8}$$

suitably large N, one clearly has $V^*[x_0, t_0; N] > -\infty$ for all x_0. Obviously
$x_0, t_0; N] < \infty$. So conditions (a) through (d) have been shown to be equivalent.
It remains to verify (4.5). From (4.7) and (4.8), it is clear that
$V^*[x_0, t_0; N] = V^*[x_0, \eta_f = 0]$. Further

$$V_\varepsilon^*[x_0, t_0] = \min_{||\eta_f|| \leq \varepsilon} [2x_0'P_{0f}\eta_f + \eta_f'P_{ff}\eta_f] + x_0'P_{00}x_0 .$$

s clear that

$$\lim_{\varepsilon \to 0} \min_{||\eta_f|| \leq \varepsilon} [2x_0'P_{0f}\eta_f + \eta_f'P_{ff}\eta_f] = 0$$

e x_0, P_{0f} and P_{ff} are fixed during the minimization and limiting operations.

$$\lim_{\varepsilon \to 0} V_\varepsilon^*[x_0, t_0] = x_0'P_{00}x_0 = V^*[x_0, \eta_f = 0]$$

required.

rks 4.2: 1. Condition (c) can be replaced by

(c') $V_\varepsilon^*[x_0, t_0]$ exists for all x_0 and all $\varepsilon > 0$

some minor calculations will show.

2. Perhaps a little more surprisingly, condition (c) can also be replaced

(c'') $V_\varepsilon^*[x_0, t_0]$ exists for all x_0 and some $\varepsilon > 0$.

reason for this is that, by a simple scaling argument, one has $V_{k\varepsilon}^*[kx_0, t_0]$
$V_\varepsilon^*[x_0, t_0]$ for all k.
In Section 3, we showed the equivalence of various conditions involving some kind
robustness with a single condition involving a Riemann-Stieltjes integral. At this
t, we cannot quite do this; as we argue below, the sort of robustness studied is
quite adequate.
One can informally think of the problem of minimizing (2.2) subject to $E_f x(t_f) = 0$
ne of minimizing the sum of the integral term in (2.2) and a terminal weighting

$x'(t_f)[S + (+\infty))E_f'E_f]x(t_f)$. Then conditions (a) and (b) of Theorem 4.2 are seen t‹ involve the replacement of this weighting term by $x'(t_f)[S + NE_fE_f]x(t_f)$ with N suitably large. In this way, some perturbation of the weighting matrix is being allo‹ but, if E_f has fewer rows than $x(\cdot)$, it is clear that part of the weighting matri‹ is not perturbed. It is then reasonable to postulate a form of robustness in which is replaced by $\bar{S}$ for some symmetric $\bar{S}$ with $||\bar{S} - S|| < \eta$ for some small positiv‹ η. Call this S-perturbation. The introduction of this type of robustness allows a connection of the ideas of Theorems 4.1 and 4.2.

Theorem II.4.3: The following conditions are equivalent.

(a) The controllability condition (4.2) holds with t replaced by t_0, and fo‹ some $\eta > 0$ and with S replaced by any symmetric $\bar{S}$ with $||\bar{S} - S|| < \eta$, $V^*[x_0, t_0; N]$ exists for some N and all x_0.

(b) For some $\eta > 0$ and any $\bar{S}$ with $||\bar{S} - S|| < \eta$, there exists on $[t_0, t_f$ a symmetric $P(\cdot)$ of bounded variation such that $dM(P) \geq 0$ within $[t_0, t_f]$ and, with Z as in Theorem 4.1,

$$Z'[P(t_f) - \bar{S}] Z \leq 0. \tag{4.9}$$

(c) An obvious modification of any of conditions (b) - (d) of Theorem 4.2 hold‹

Proof: (a) $\Longleftrightarrow$ (c) is trivial. We show first that (a) $\Longrightarrow$ (b). By Theorem 3.3, there exists a symmetric $P(\cdot)$ of bounded variation on $[t_0, t_f]$, with $dM(P) \geq 0$ within $[t_0, t_f]$ and with

$$P(t_f) - \bar{S} - NE_f'E_f \leq 0.$$

Equation (4.9) is immediate. To show (b) $\Longrightarrow$ (a), observe that with $\bar{\eta} < \eta - ||\bar{S} - S||$ one has $||\bar{S} - \bar{\eta}I - S|| < \eta$ and so, using condition (b) with $\bar{S}$ replaced by $\bar{S} - \bar{\eta}$ there exists $P(\cdot)$ with the various properties stated except that (4.1) is replaced by

$$Z'[P(t_f) - \bar{S} + \bar{\eta}I]Z \leq 0$$

whence

$$Z'[P(t_f) - \bar{S}]Z < 0.$$

Some matrix algebra will then show that there exists N for which

$$\begin{bmatrix} Z' \\ E_f \end{bmatrix} [P(t_f) - \bar{S} - NE_f'E_f][Z \quad E_f'] < 0$$

or

$$P(t_f) - \bar{S} - NE_f'E_f < 0.$$

Theorem 3.3 then implies condition (a).

<u>Remarks 4.3</u>: 1. The crucial step in the above proof is to replace (4.9) by a strict inequality. This is not possible without the S-perturbation idea.

2. In case E_f is square, there is no need for the S-perturbation idea at all. Conditions (4.4) and (4.9) evanesce.

Now we examine the third type of robustness – that involving perturbation of the terminal state t_f. To simplify the results, we shall consider constraints of the type $x(t_f) = 0$.

<u>Theorem II.4.4</u>: Suppose that the controllability condition (4.2) holds. Then the following conditions are equivalent.

(a) $\overset{*}{v}[x_0, t_0; N]$ exists for some N and all x_0.

(b) There exists $t_1 > t_f$ with definitions of $F(\cdot)$, $G(\cdot)$, etc. on $(t_f, t_1]$ such that these quantities are continuous on $[t_0, t_1]$, such that

$$\int_t^{t_1} \Phi(t_1, \tau)G(\tau)G'(\tau)\Phi'(t_1, \tau)d\tau > 0$$

for all $t \in [t_0, t_1)$ and such that the constrained end-point problem has a solution for all $x(t)$ and $t \in [t_0, t_1)$.

Proof: (a) $\Rightarrow$ (b). By Theorem 3.3, there exists a symmetric $P(\cdot)$ of bounded variation on $[t_0, t_f]$ satisfying $dM(P) \geq 0$ within $[t_0, t_f]$ and $P(t_f) \leq S + NI$. Apply a minor modification of the extension procedure of the proof of Theorem 3.4 to define $F(\cdot)$, $G(\cdot)$, etc. with the properties stated and $P(\cdot)$ such that $dM(P) \geq 0$ within $[t_0, t_1]$. By Theorem 4.1, the constrained end-point problem defined over $[t, t_1]$ has a solution.

(b) $\Rightarrow$ (a). Suppose that

$$x'(t)P(t)x(t)=\inf_{u(\cdot)} \int_t^{t_1} [x'Q(t)x + 2u'H(t)x + u'R(t)u]dt$$

subject to $x(t_1) = 0$. Then for $t \leq t_f$ and N such that $NI \geq P(t_f)$ we obtain a lower bound for $\overset{*}{v}[x_0, t_0; N]$:

$$x'(t_0)P(t_0)x(t_0) = \inf_{u(\cdot)} \{\int_{t_0}^{t_f}[x'Q(t)x + 2u'H(t)x + u'R(t)u]dt$$
$$+ x'(t_f)P(t_f)x(t_f)\}$$
$$\leq \inf_{u(\cdot)} \{\int_{t_0}^{t_f}[x'Q(t)x + 2u'H(t)x + u'R(t)u]dt$$
$$+ N||x(t_f)||^2\}$$
$$= \overset{*}{v}[x_0, t_0; N].$$

This proves the result.

Remark 4.4: Consideration of constraints of the type $E_f x(t_f) = 0$ for nonsquare E_f is more messy. It does not seem possible to obtain tidy results for which the extended interval constraint becomes $E_f x(t_1) = 0$; rather, one should consider $E_f \Phi(t_f, t_1) x(t_1) = 0$ and now $\Phi(\cdot, \cdot)$ depends on the particular $F(\cdot)$ chosen.

5. EXTREMAL SOLUTIONS OF RIEMANN-STIELTJES INEQUALITIES

In this section, we study the inequality $dM(P) \geq 0$ within $[t_0, t_f]$ and determine maximum and minimum solutions of the inequality. To begin with, we have the following result:

Theorem II.5.1: Suppose in relation to the unconstrained minimization problem defined by (3.1) and (3.2) [$x(t_0)$ fixed but arbitrary and $x(t_f)$ free] that, for some terminal weighting matrix S_f, $V^*[x(t), t] = x'(t) P^*(t) x(t)$ exists for all $x(t)$ and $t \in [t_0, t_f]$. Then for any $P(\cdot)$ such that $dM(P) \geq 0$ and $P(t_f) \leq S_f$, we have

$$P(t; \; P(t_f) \leq S_f) \leq P^*(t; \; P^*(t_f) = S_f)$$

$$\forall \; t \in [t_0, t_f]. \tag{5.1}$$

(The notation should be self explanatory; though perhaps needlessly complicated at this point, it will be helpful later).

Proof: From the proof of Theorem 3.2 we know that for $t \in [t_0, t_f]$,

$$\int_t^{t_f} [x'(t) \; u'(t)] dM(P) \begin{bmatrix} x(t) \\ u(t) \end{bmatrix} = x'(t_f) P(t_f) x(t_f) - x'(t) P(t) x(t)$$

$$+ \int_t^{t_f} [x'Q(\tau)x + 2x'H(\tau)u + u'R(\tau)u] \, d\tau.$$

Therefore

$$V[x(t), t, u(\cdot)] = x'(t) P(t) x(t) + x'(t_f) [S_f - P(t_f)] x(t_f)$$

$$+ \int_t^{t_f} [x'(t) \; u'(t)] dM(P) \begin{bmatrix} x(t) \\ u(t) \end{bmatrix}.$$

Using the conditions on $P(\cdot)$, it follows that

$$V[x(t), t, u(\cdot)] \geq x'(t) P(t) x(t).$$

Equation (5.1) is immediate.

As we know, $P^*(\cdot)$ satisfies $dM(P^*) \geq 0$ and $P^*(t_f) \leq S_f$; thus $P^*(\cdot)$ is the maximal $P(\cdot)$ with this property, the ordering being defined by (5.1).

We obtain a minimal $P_*(\cdot)$ in the following way.

Theorem II.5.2: Define the performance index

$$V_0[x(t), t, u(\cdot)] = \int_{t_0}^{t} [x´Qx + 2x´Hu + u´Ru]dt - x´(t_0)S_0x(t_0) \quad (5.2)$$

in which $x(t)$ is arbitrary but fixed, $x(t_0)$ is free, and $u(\cdot)$ is free. Set

$$V_*[x(t), t] = \inf_{u(\cdot)} V_0[x(t), t, u(\cdot)] . \quad (5.3)$$

If $V_*[x(t), t]$ exists for all $x(t)$ and $t \in [t_0, t_f]$, then $V_*[x(t), t] = -x´(t)P_*(t)x(t)$ for some $P_*(t)$ defined on $[t_0, t_f]$. Moreover, there exists $P(\cdot)$ such that $dM(P) \geq 0$ within $[t_0, t_f]$ and $P(t_0) \geq S_0$, and $P_*(\cdot)$ is the minimal such $P(\cdot)$:

$$P(t; \ P(t_0) \geq S \ ) \geq P_*(t; \ P_*(t_0) = S_0)$$

$$t \geq [t_0, t_f] \quad (5.4)$$

Conversely, if there exists $P(\cdot)$ such that $dM(P) \geq 0$ with $[t_0, t_f]$ and $P(t_0) \geq S$, then $V_*[x(t), t]$ exists for all $x(t)$ and $t \in [t_0, t_f]$.

The proof of this theorem can be obtained by time reversal of Theorem 5.1, and Theorem 3.3, which relates the existence of V^* to the existence of $P(\cdot)$ satisfying $dM(P) \geq 0$ within $[t_0, t_f]$ and $P(t_f) \leq S_f$.

Theorem II.5.2 now provides information about those terminal weighting matrices for which $V^*[x(t), t]$ exists for all $x(t)$ and $t \in [t_0, t_f]$. In fact, the theorem was introduced not for its intrinsic content, but in order to provide this information.

Theorem II.5.3: With notation as above, let $V_*[x(t), t]$ exist for all $x(t)$, all $t \in [t_0, t_f]$, and for some S_0. Then it exists for all $\bar{S}_0 \leq S_0$. Moreover, $V^*[x(t), t]$ exists for all S_f with

$$S_f \geq P_*(t_f; \ P_*(t_0) = \bar{S}_0) \quad (5.5)$$

for some $\bar{S}_0 \leq S_0$. If (5.5) fails for all $\bar{S}_0 \leq S_0$ and some S_f, then $V^*[x(t), t]$ does not exist for all $x(t)$ and $t \in [t_0, t_f]$.

Proof: From (5.2), we see that $V[x(t), t]$ increases as S_0 decreases. Therefore $V_*[x(t), t] = -x´(t)P_*(t)x(t)$ increases as S_0 decreases. Hence if $V_*[x(t), t]$ exists for some S_0, it exists for all $\bar{S}_0 \leq S_0$. [Moreover, we see that $P_*(t; \ P_*(t_0) = \bar{S}_0)$ for each fixed t is monotone with S_0]. This proves the first claim.

To prove the second claim, suppose that (5.5) holds for some $\bar{S}$. Then $P(t) = P_*(t; P_*(t_0) = \bar{S}_0)$ is such that $dM(P) \geq 0$ and $P(t_f) \leq S_f$. By Theorem 3.3, $V^*[x(t), t]$ exists for all $x(t)$ and $t \in [t_0, t_f]$.

Conversely, suppose that (5.5) fails for some S_f and for all $\bar{S}_0 \leq S_0$, but that $V^*[x(t), t]$ exists for all $x(t)$ and $t \in [t_0, t_f]$. We shall obtain a contradiction. Let $\hat{S}_0 = P^*(t_0)$. Then since $dM(P^*) \geq 0$ within $[t_0, t_f]$ and $P^*(t_0) \geq \hat{S}_0$ we have by Theorem 5.2 that $V_*[x(t), t]$ exists for all t and $x(t) \in [t_0, t_f]$, with S_0 replaced by $\hat{S}_0$ in (5.2). Moreover, from (5.4),

$$P^*(t) \geq P_*(t; P_*(t_0) = \hat{S}_0)$$

for all $t \in [t_0, t_f]$. Taking $t = t_f$ yields that $S_f \geq P_*(t_f; P_*(t_0) = \hat{S}_0)$. Now for $\bar{S}_0$ suitably negative, $\bar{S}_0 \leq \hat{S}_0$ and $\bar{S}_0 \leq S_0$. Then $P_*(t_f; P_*(t_0) = \bar{S}_0) \leq P^*(t_f; P_*(t_0) = \hat{S}_0) \leq S_f$, which is a contradiction. This proves the theorem.

Remark 5.1:   1.   In view of the monotonicity of $P_*(t; P_*(t_0) = S_0)$ with S_0, we see that any S_f for which $S_f \geq P_*(t_f; P_*(t_0) = -nI)$ for some arbitrarily large n is such that the corresponding $V^*[x(t), t]$ exists on $[t_0, t_f]$. Likewise, any S_0 for which $S_0 \leq P^*(t_0; P^*(t_f) = nI)$ for arbitrarily large n is such that the corresponding $V_*[x(t), t]$ exists on $[t_0, t_f]$.

2.   It is possible to obtain results for constrained minimization problems in which part or all of the state vector is constrained at t_0, t_f. The most interesting of these is one that follows easily from the three theorems above:

$$P_*(t; x(t_0) = 0) \leq P(t) \leq P^*(t; x(t_f) = 0) \qquad (5.6)$$

In this inequality, it is assumed that the requisite controllability conditions are satisfied and that the first and third quantities are well-defined on $(t_0, t_f]$ and $[t_0, t_f)$ respectively. The quantity $P(t)$ is any solution to $dM(P) \geq 0$ on $[t_0, t_f]$. The inequality is reminiscent of some known for time-invariant problems, see [15]. The bulk of the results of this section first appeared in [16].

6.   SUMMARIZING REMARKS

The main thrust of the chapter has been to show that there exist conditions involving the nonnegativity of certain Riemann-Stieltjes integrals which are both necessary and sufficient for certain linear-quadratic optimization problems to have a solution. These problems are not identical, though they are closely related, to those conventionally examined in the literature; rather, they have an inherent quality of robustness, which makes them qualitatively well-posed.

One set of results relate to robustness around the initial time or state, and a second set to robustness round a final time or constrained state. In the latter context, we have shown that penalty function ideas can be employed, and, moreover, robust lems are the only class of problems to which they can be applied.

On the grounds then of mathematical tidiness and the rational appear of qualitat-
y well-posed problems, we suggest a change of viewpoint as to which linear-
catic minimization problems should be thought of as standard.

We also have pointed out the applicability of Riemann-Stieltjes type conditions
urther linear-quadratic control problems, including those requiring transfer from
nitial state of zero to a prescribed nonzero terminal state. The extension is
to achieve, by a time-reversal argument. It is then possible to characterize
cal properties of extremal solutions of inequalities involving Riemann-Stieltjes
grals, in the process linking various classes of problems whose analysis in terms
uch integrals is feasible.

RENCES

J.B. Moore and B.D.O. Anderson, "Extensions of quadratic minimization theory, I:
Finite time results", *Int. J. Control*, Vol. 7, No. 5, 1968, pp. 465-472.

D.H. Jacobson, "Totally singular quadratic minimization problems", *IEEE Trans.
Automatic Control*, Vol. AC-16, 1971, pp. 651-658.

B.D.O. Anderson, "Partially singular linear-quadratic control problems", *IEEE
Trans. Automatic Control*, Vol. AC-18, 1973, pp. 407-409.

B.P. Molinari, "Nonnegativity of a quadratic functional", *SIAM J. Control*, Vol. 13,
1975, pp. 792-806.

W.A. Coppel, "Linear-quadratic optimal control, *Proc. Roy. Soc. Edin.*, Vol. 73A,
1974-5, pp. 271-289.

D.J. Clements, B.D.O. Anderson and P.J. Moylan, "Matrix inequality solution to
linear-quadratic singular control problems", *IEEE Trans. Automatic Control*, Vol.
AC-22, 1977, pp. 55-57.

D.H. Jacobson and J.L. Speyer, "Necessary and sufficient conditions for singular
control problems: a limit approach", *J. Math. Anal. Appl.*, Vol. 34, 1971,
pp. 239-266.

R.W. Brockett, Finite Dimensional Linear Systems, John Wiley, New York, 1970.

P.A. Faurre, "Sur les points conjugués en commande optimale", *C.R. Acad. Sci.*,
Ser. A, Vol. 266, 1968, pp. 1294-1296.

I.M. Gelfand and S.V. Fomin, Calculus of Variations, Prentice-Hall, N.J., 1963.

B.D.O. Anderson and P.J. Moylan, "Spectral factorization of a finite-dimensional
nonstationary matrix covariance", *IEEE Trans. Automatic Control*, Vol. AC-19, 1974,
pp. 680-692.

B.D.O. Anderson and P.J. Moylan, "Synthesis of linear time-varying passive networks",
IEEE Trans. Circuits and Systems, Vol. CAS-21, 1974, pp. 678-687.

D.G. Luenberger, Introduction to Linear and Nonlinear Programming, Addison-
Wesley Publishing Co. Inc., London, 1973.

B.D.O. Anderson and D.J. Clements, "Robust linear-quadratic minimization", *J.
Math. Anal. Appl.*, to be published.

[15] J.C. Willems, "Least squares stationary control and the algebraic Riccati equation", *IEEE Trans. Automatic Control*, Vol. AC-16, 1971, pp. 621-234.

[16] D.J. Clements and B.D.O. Anderson, "Extremal solutions of Riemann-Stieltjes inequalities of linear optimal control", *IEEE Trans. Automatic Control*, Vol. AC-22, 1976, pp. 139-140.

CHAPTER III

LINEAR-QUADRATIC SINGULAR CONTROL: ALGORITHMS

INTRODUCTION

In this chapter we are concerned with the existence and computation of optimal ntrols in the singular, linear-quadratic control problem with no end-point constraints. this end, we initially look at the slightly simpler problem of finding necessary and fficient conditions for a quadratic cost functional to be bounded below independently the control function, subject to linear differential equation constraints.

We are again interested in the system and cost defined by equations (II.3.1) and I.3.2). For convenience, we rewrite these equations here. That is, we consider the st

$$V[x_0, t_0, u(\cdot)] = \int_{t_0}^{t_f} \{x'Q(t)x + 2x'H(t)u + u'R(t)u\}dt$$
$$+ x'(t_f)Sx(t_f) \tag{1.1}$$

d dynamics

$$\dot{x} = F(t)x + G(t)u, \quad x(t_0) = x_0. \tag{1.2}$$

We make the same assumptions on the coefficient matrices and controls as in the evious chapter. However, we <u>further</u> assume that the various matrices F, G, etc., l have differentiability properties sufficient to allow the carrying out of certain ansformations (involving differentiation) which are explained subsequently. The mber of such transformations can vary from problem to problem and so consequently es the required degree of differentiability of the coefficient matrices. At each ate of the development of the algorithm described in this chapter we shall state the gree of differentiability sufficient for the carrying out of that stage. (In case ly one complete cycle of the algorithm is required, continuous differentiability of R, F and G, and H twice continuously differentiable are sufficient.) sides this class of assumptions, we shall also on occasions need to assume <u>further</u> e constancy of rank on $[t_0, t_f]$ of certain matrices constructed from F, G, etc.

Denote the set of admissible controls by U. As in Chapter II, we are interested the problem of finding necessary and sufficient conditions for

$$V[0, t_0, u(\cdot)] \geq 0 \quad \text{for each} \quad u(\cdot) \in U \tag{1.3}$$

bject to (1.2). In case R(t) > 0 for all t ∈ $[t_0, t_f]$ i.e. the problem is non-ngular, it is easily solved. The interesting cases are those when R(t) ≡ 0 (the tally singular case), and R(t) is nonzero but singular somewhere in $[t_0, t_f]$ he partially singular case).

We again stress the fact that this problem (1.3) has already been looked at in Chapter II where we were primarily interested in general existence conditions. Here, our interest is in how we might determine, by construction, whether or not (1.3) holds. To do this, we need to study a more restricted (in the sense that differentiability and constancy of rank conditions need to hold) version of the problem than that studied in Chapter II.

Historically this problem has proved to be important in several areas. It is the second variation problem of optimal control [1], and is closely connected with the linear-quadratic control problem [2], [3]. It also appears in the dual control problem of covariance factorization [4], [5], [6], and finally, one definition of passivity leads to a similar problem in network synthesis [7].

Initially, however, it was as the second variation problem of optimal control that the question was studied. Stronger necessary conditions than the classical Legendre-Clebsch condition were needed to eliminate singular extremals from consideration as minimizing arcs for problems which arose in aerospace trajectory optimization. For more detailed information of the history of this problem see the surveys [8], [9] and the book [10], together with the references therein. Arising from these studies were the generalized Legendre-Clebsch conditions which in the totally singular case can be written

$$\frac{\partial}{\partial u} \frac{d}{dt} \left(\frac{\partial H}{\partial u} \right)' = 0 \qquad \text{on } [t_0, t_f] \tag{1.4}$$

$$-\frac{\partial}{\partial u} \frac{d^2}{dt^2} \left(\frac{\partial H}{\partial u} \right)' \geq 0 \qquad \text{on } [t_0, t_f] \tag{1.5}$$

where H is the Hamiltonian associated with (1.1) and (1.2), i.e.

$$H = x'Q(t)x + 2x'H(t)u + \lambda'(F(t)x + G(t)u) \tag{1.6}$$

$$-\dot{\lambda} = \frac{\partial H}{\partial x}$$

and λ is the costate vector. If (1.5) is met with equality, the procedure leading to (1.4) and (1.5) can be extended to give further necessary conditions. In general, the necessary conditions become

$$\frac{\partial}{\partial u} \frac{d^q}{dt^q} \left(\frac{\partial H}{\partial u} \right)' = 0 \qquad \text{on } [t_0, t_f], \quad q < 2p \tag{1.8}$$

$$(-1)^p \frac{\partial}{\partial u} \frac{d^{2p}}{dt^{2p}} \left(\frac{\partial H}{\partial u} \right)' \geq 0 \qquad \text{but nonzero on } [t_0, t_f] \tag{1.9}$$

where $\frac{d^{2p}}{dt^{2p}} \left(\frac{\partial H}{\partial u} \right)'$ is the lowest order time derivative of $\left(\frac{\partial H}{\partial u} \right)'$ in which some component of the control u appears explicitly with a nonzero coefficient. The integer p is called the order of the singular arc for scalar u; for vector u an extension of this definition is needed.

These conditions (1.8), (1.9) were initially derived by Kelley [12], [13] for

controls only, in which case it is not difficult to show that, for odd q,   (1.8)
omatically satisfied. The original derivation [12] used the classical method of
ucting special variations and considering terms of comparable orders. In [13]
4], a transformation technique for deriving (1.9) is described and it is this
ormation which will be studied here for the general case of vector controls.
t should be noted that passing from the scalar case to the vector case is gener-
ar from easy. To suggest why the extension is nontrivial, (1.5) can be examined.
 scalar case, two possibilities arise, of equality [leading on to (1.8) and (1.9)]
quality. In the vector control case, there are really three possibilities;   (1.5)
ld with equality [leading as before to (1.8) and (1.9), or at least some of these
ons] or it can hold with strict inequality (as with the scalar case), or it can
ith a loose inequality, the matrix on the left side of (1.5) being singular and
ɔ. Some modification of (1.8) and (1.9) is called for to cope with this case.
 dual problem of spectral factorization, the vector problem though now solved,
uch longer to solve than the scalar problem; this fact also suggests the non-
lity of the scalar-to-vector extension. Nevertheless results have been obtained
e vector control problem; a general form of the generalized Legendre–Clebsch
ions [(1.8) and (1.9) being inadequate to cover all possibilities as just noted]
en derived by Robbins [11] and Goh [15]. Robbins' method was essentially variat-
 whereas Goh used a transformation of the states and controls in a treatment
represents an application of work he had done on the singular Bolza problem in
lculus of variations [16]. Though both Kelley and Goh used transformation
s, there is a major difference in the style of the transformations. Kelley's
ormation procedure replaces the original performance index and linear system
on by one involving a state variable of lower dimension than the original. Goh
s the full state-space dimension and there arise as a result a number of extra
aint conditions over and above those which might fairly be termed generalized
re-Clebsch conditions. These extra constraint conditions have been examined at
 in [17].
s noted above, in this chapter we are interested in a set of necessary and suff-
 conditions for (1.3) to hold subject to (1.2), of more limited applicability
he necessary conditions and sufficient conditions of Theorems II.3.1 and II.3.2.
uitation stems from the need to have certain differentiability of constancy of
onditions satisfied; the advantage gained is that the conditions are highly
ent to the problem of computing an optimal control and performance index for
r (1.1), (1.2) with free $x(t_0)$.
ese conditions fit in with previous work in the following way. First, they
 extension of conditions published in [18] applicable to scalar controls when
gular arcs are of order 1. Second, the conditions are obtained using a vector
lization of the Kelley transformation (which, it should be recalled, is limited
 scalar control case). Third, the various steps required to obtain the con-

ditions are in large measure the dual of those arising in an algorithm of Anderson
and Moylan, used in [5] and [7] for non-control purposes. We discuss the algorithm a
little further.

The problem of constructing a P matrix that appears in Riemann-Stieltjes inequalities as described in the previous chapter is central to the problems of covariance
factorization and time-varying passive network synthesis. It was in the latter context that an algorithm suitable for the stationary case was developed [19], and then
it was recognized that this algorithm with variations was also applicable to the time-
varying synthesis problem [7], and with other variations to the covariance factorization problem [5]. An algorithm was in fact suggested in [5] for finding such a P
matrix under additional differentiability and constancy of rank assumptions. In this
work, we show that the Anderson-Moylan algorithm is precisely (the vector extension of)
Kelley's transformation executed in a particular co-ordinate basis and in showing this,
we derive the generalized Legendre-Clebsch conditions in a reasonably straightforward
manner.

In connection with the optimal control problem associated with free $x(t_0)$, the
Anderson-Moylan algorithm, considered in isolation from the Kelley transformation
procedure, can be shown to yield the optimal performance index. Linking it with the
Kelley transformation procedure yields the optimal control as well.

Finally we mention the work in [20-23]. In [20], the solution of the singular
regulator problem is studied by obtaining the asymptotic solution of the regularized
problem (see comments prior to Theorem II.3.2) as $\varepsilon \to 0$. The methods used are those
of singular perturbation theory of ordinary differential equations. The work in [21],
of which we only became aware after completing the studies in this chapter, partially
raises some of the problems raised here but is nowhere near as complete. References
[22, 23] contain much of the work of this chapter.

We now outline the structure of this chapter. Section 2 is concerned with develop-
ing a standard form for the control problem and, in the process, possibly reducing the
control space dimension. Section 3 develops the general Kelley transformation for the
nonnegativity problem in standard form, producing the generalized Legendre-Clebsch
conditions and a set each of necessary and sufficient conditions for the existence of
a solution to the nonnegativity problem. In the process, the state-space dimension
is reduced. In Section 4, the results of Sections 2 and 3 are applied to the linear-
quadratic optimal control problem. Using a sequence of control and/or state-space
dimension reductions, minimizing (or infimizing) controls and the corresponding optimal
cost are calculated. The results in Section 5 link an algorithm used in the dual
problem of covariance factorization with the algorithm outlined in Section 4; here,
heavy use is made of the Riemann-Stieltjes inequality of the preceding chapter.
Section 6 contains summarizing remarks.

CONTROL SPACE DIMENSION REDUCTION AND A STANDARD FORM

In this section, our aim is to show how extraneous controls may be removed, and
, after their removal, a certain standard form may be assumed for the matrices   R
 G. All this is done with the aid of coordinate basis changes of the input and
.te spaces; differentiability and constancy of rank assumptions need to be invoked.
We begin with some preliminary and simple observations.

The problem of minimizing (1.1) subject to (1.2) with initial condition $x(t_0)$
is equivalent to the problem of minimizing (1.1) subject to (1.2) with initial
condition $x(t_0)$ and with $u(t)$, $R(t)$, $H(t)$ and $G(t)$ replaced by $\bar{u}(t)$
$= U^{-1}(t)u(t)$, $\bar{R}(t) = U'(t)R(t)U(t)$, $\bar{H}(t) = H(t)U(t)$ and $\bar{G}(t) = G(t)U(t)$ for
any nonsingular matrix $U(t)$, continuous on $[t_0, t_f]$. This statement corresponds to a change of basis of the control space.

The problem of minimizing (1.1) subject to (1.2) with initial condition $x(t_0)$
is equivalent to the problem of minimizing (1.1) subject to (1.2) with initial
condition $U(t_0)x(t_0)$ and with $x(t)$, $F(t)$, $H(t)$, $G(t)$, $Q(t)$ and S replaced
by $\bar{x}(t) = U(t)x(t)$, $\bar{F}(t) = U(t)F(t)U^{-1}(t) + \dot{U}(t)U^{-1}(t)$, $\bar{G}(t) = U(t)G(t)$,
$\bar{H}(t) = [U^{-1}(t)]'H(t)$, $\bar{Q}(t) = [U^{-1}(t)]'Q(t)U^{-1}(t)$ and $\bar{S} = [U^{-1}(T)]'SU^{-1}(T)$ for
any nonsingular matrix $U(t)$, continuously differentiable on $[t_0, t_f]$. This
statement corresponds to a change of basis of the state space. Note that in
observation 1 continuity of $U(t)$ is sufficient whereas in observation 2 the
stronger condition of continuous differentiability is required.

The transformation procedure now follows.

p 1.
Assumption 1: $R(t)$ has constant rank r on $[t_0, t_f]$.
With this assumption, an application of Dolezal's Theorem, see Appendix A, guaran-
s the existence of a matrix $U(t)$, nonsingular and continuous on $[t_0, t_f]$ such
t

$$\bar{R}(t) \underset{=}{\triangle} U'(t)R(t)U(t) = \begin{bmatrix} I_r & 0 \\ 0 & 0 \end{bmatrix}. \qquad (2.1)$$

 this $U(t)$ we change the basis of the control space.

p 2.
Partition G as $[G_1 \quad G_2]$ where G_1 is an n×r matrix.
Assumption 2: $G_2(t)$ has constant rank s ≤ m-r. If s = m-r pass to Step 3.
erwise, another application of Dolezal's Theorem implies the existence of a matrix
t), nonsingular and continuous on $[t_0, t_f]$ such that $G_2(t)V_0(t) = [\hat{G}_2(t) \quad 0]$
 $\hat{G}_2(t)$ having s columns. Set $V(t) = I \oplus V_0(t)$, with the unit matrix of
ension r. With this $V(t)$ we change the basis of the control space. Having done

this we partition R, G and H as

$$R(t) = \begin{bmatrix} I_r & 0 \\ 0 & 0_{(m-r)\times(m-r)} \end{bmatrix}$$

$$G(t) = [G_1(t) \quad \hat{G}_2(t) \quad 0_{n\times(m-r-s)}]$$

$$H(t) = [H_1(t) \quad H_2(t) \quad H_3(t)].$$

Writing $u'(t) = [\hat{u}'(t) \quad u_3'(t)]$ where $\hat{u}(t)$ has dimension r+s, one obtains from (1.1) and (1.2),

$$V[x_0, t_0, u(\cdot)] = x'(t_f)Sx(t_f) + \int_{t_0}^{t_f}\{x'Qx + 2x'\hat{H}\hat{u} + \hat{u}'\hat{R}\hat{u}\}dt + 2\int_{t_0}^{t_f}x'H_3u_3dt \qquad (2.2)$$

$$\dot{x} = Fx + \hat{\hat{G}}\hat{u} \qquad (2.3)$$

where

$$\hat{R}(t) = \begin{bmatrix} I_r & 0 \\ 0 & 0_{s\times s} \end{bmatrix}$$

$$\hat{G}(t) = [G_1(t) \quad \hat{G}_2(t)]$$

$$\hat{H}(t) = [H_1(t) \quad H_2(t)].$$

We now focus on the last term in (2.2):

<u>Lemma III.2.1</u>: A necessary condition for $V[0, t_0, u(\cdot)]$ to be nonnegative for all $u(\cdot)$ is that $\xi'(t)H_3(t) = 0$ for all states $\xi(t)$ reachable from $x(t_0) = 0$ for $t \in [t_0, t_f]$. A necessary condition for $V^*[x_0, t_0]$ to be finite for all x_0 is that $H_3(t) = 0$ for $t \in [t_0, t_f]$.

<u>Proof</u>: Assume $t < t_f$. Let $\hat{u}_\xi$ on $[t_0, t]$ take $x(t_0) = 0$ to $x(t) = \xi(t)$. Set $\hat{u}_\xi \equiv 0$ on $[t, t_f]$ and $u_3 = v$ on $[t, t+\epsilon]$ for small $\epsilon > 0$ and zero elsewhere. Then

$$V[0, t_0, u(\cdot)] = \text{constant} + 2\epsilon\xi'(t)H_3(t)v + \text{terms of higher order in}\ \epsilon.$$

Unless $\xi'(t)H_3(t) = 0$, we can readily obtain a contradiction to the nonnegativity of $V[0, t_0, u(\cdot)]$ by appropriate choice of v. The second part of the lemma follows from the observation that at time t, all states are reachable from some x_0. Therefore $H_3(t) = 0$ on $[t_0, t_f)$, and therefore $[t_0, t_f]$ by continuity.

If the condition expressed in the lemma fails, no further computations are needed to check the nonnegativity or finite infimum condition. If the condition holds, then

.n checking for nonnegativity there is no loss of generality in assuming $H_3(t) \equiv 0$ ɔn $[t_0, t_f]$ (since the performance index is unaltered). Then the u_3 component of ı can be dispensed with, and, dropping the hat superscript, we obtain a problem of the same form as the given one, but with lower control space dimension. It should be clear that in the context of linear-quadratic control problems this step corresponds to the throwing away of those controls which have no effect on the states via (2.3) and do not appear directly in (2.2).

Step 3.

Using the fact that $\hat{G}_2(t)$ has s columns and rank s, Dolezal's theorem again guarantees the existence of a nonsingular matrix $T_0(t)$, continuous on $[t_0, t_f]$ such that $T_0(t)\hat{G}_2(t) = [0^{'} \quad I_{s \times s}]^{'}$. Set $T(t) = I_r \oplus T_0(t)$. Now continuity of $T(t)$ is not sufficient for a state space change of basis. Dolezal's theorem guarantees the same degree of differentiability for $T_0(t)$ as the matrix $G_2(t)$; therefore we make the

Assumption 3: G (t) has continuously differentiable elements.

With T (t), we then change the basis of the state space.

The end result of these three steps, depending for their execution on Assumptions 1-3 and on a nonnegativity or finite infimum assumption, is that

$$ R = \begin{bmatrix} I_r & 0 \\ 0 & 0 \end{bmatrix} \qquad G = \begin{bmatrix} G_{11} & 0 \\ G_{21} & I_s \end{bmatrix} \qquad\qquad (2.4) $$

where G_{21} has r columns. Notice that there is nothing special about the other matrices defining the problem. Notice also that, (2.4) results whether or not some of the controls are eliminated.

When R and G are as in (2.4), we shall say that the control problem is in standard form. There is then a natural partitioning of the state vector x as $[x_1^{'} \quad x_2^{'}]^{'}$ where x_2 is s-dimensional, and the control vector u as $[u_1^{'} \quad u_2^{'}]^{'}$ where u_2 is s-dimensional. We call u_1 and u_2 the nonsingular and singular controls respectively; this nomenclature arises because given the form of R in (2.4), the quadratic term $u_1^{'}u_1$ occurs in the performance index while the vector u_2 occurs at most linearly through the term $x^{'}Hu$. Moreover, the singular controls all independently influence the state x_2 [as is clear from (2.4)] and hence the cost functional (1.1).

3.   VECTOR VERSION OF KELLEY TRANSFORMATION

In this section, we shall extend Kelley's transformation to the vector case, taking our problem in the standard form derived in the last section. We shall deduce the generalized Legendre-Clebsch conditions from the transformed problem, and shall then show that the application of Kelley's transformation to our problem leads to a

set of necessary and sufficient conditions for the solution of our problem. These
consist of the existence of a solution to a problem of the same form but of lower state
dimension, a set of end-point constraints and the relevant generalized Legendre-Clebsch
condition corresponding to equation (1.4).

The development of this section will essentially follow that of [14] and [18],
generalized to the vector case as in [22, 23]. Assume that we are given (1.1) and (1.2)
with R and G given by (2.4), and that we are interested in finding necessary conditions for (1.3) to hold subject to (1.2). Construct the Mayer form of the problem
by introducing the scalar variable w_0 defined by

$$\dot{w}_0 = x'Qx + 2x'Hu + u'Ru \quad , \quad w_0(t_0) = 0 \; . \tag{3.1}$$

Equations (1.2) and (3.1) now define a set of $(n+1)$ differential equations in the
variables w_0 and x. Recalling the standard form of R and G, and the resultant
partitioning of u and x, it is clear that (3.1) involves u_2 linearly but not
quadratically. Moreover, from the partitioned form of (1.2) and (2.4), with obvious
definitions of F_{11}, etc.,

$$\dot{x}_1 = F_{11}x_1 + F_{12}x_2 + G_{11}u_1 \tag{3.2}$$

$$\dot{x}_2 = F_{21}x_1 + F_{22}x_2 + G_{21}u_1 + u_2 \tag{3.3}$$

we see that (3.2) is influenced by u_2 only indirectly via (3.3). Clearly, if (3.1)
did not contain a u_2 term at all, the original problem could intuitively be replaced
by one with state x_1 (of lower dimension than x) and controls x_2 and u_1, since
u_2 is essentially x_2 differentiated.

With this in mind, we attempt to find a transformation to new variables z_0, z_1
and z_2, with z_0 scalar and $z = [z_1' \quad z_2']'$ the corresponding partitioning of the
n-dimensional vector z, such that the dynamics of z_0 and z_1 are independent of
u_2. (Thus z_0 plays the role of the performance index, and z the role of the state
variable). Suppose that we set

$$z_0 = h_0(w_0, x_1, x_2)$$

$$z_1 = h_1(w_0, x_1, x_2) \tag{3.4}$$

$$z_2 = h_2(w_0, x_1, x_2)$$

and assume that all first order partial derivatives of h_0, h_1 and h_2 with respect
to w_0, x_1 and x_2 exist; then the dynamics of z_i, $i = 0, 1, 2$, can be written

$$\dot{z}_i = \frac{\partial h_i}{\partial w_0} \dot{w}_0 + \left(\frac{\partial h_i}{\partial x_1} \right) \dot{x}_1 + \left(\frac{\partial h_i}{\partial x_2} \right) \dot{x}_2 \tag{3.5}$$

where $\dfrac{\partial h_0}{\partial x}$ is a column vector and $\dfrac{\partial h_1}{\partial x}$ is a matrix with j-th column the gradient of
the j-th component of h_1.

Using (3.1), (3.2) and (3.3) in (3.5) for i = 0, 1, and setting the coefficients
u_2 to zero we obtain, as necessary and sufficient conditions for $\dot{z}_0$ and $\dot{z}_1$ to
ndependent of u_2.

$$\frac{\partial h_0}{\partial w_0}(2x'H_{12} + 2x_2'H_{22}) + (\frac{\partial h_0}{\partial x_2})' = 0 \tag{3.6}$$

$$\frac{\partial h_1}{\partial w_0}(2x_1'H_{12} + 2x_2'H_{22}) + (\frac{\partial h_1}{\partial x_2})' = 0 \ . \tag{3.7}$$

, we would like to find functions h_0 and h_1 satisfying the partial different-
equations (3.6) and (3.7). A standard tool for tackling such problems is the
od of base characteristics [27], which in this case proceeds as follows. Suppose
s a vector variable such that h_0 and h_1 are constant on surfaces described
arying θ, i.e.

$$\frac{\partial h_0}{\partial \theta} = 0 \qquad \text{and} \qquad \frac{\partial h_1}{\partial \theta} = 0.$$

we see from (3.6) and (3.7), that these equations will hold provided that

$$(\frac{\partial w_0}{\partial \theta})' = 2x_1'H_{12} + 2x_2'H_{22} \quad , \quad \frac{\partial x_1}{\partial \theta} = 0 \quad , \quad \frac{\partial x_2}{\partial \theta} = I.$$

ng the latter of these equations we set $\theta = x_2$ so that the remaining equations
me

$$(\frac{\partial w_0}{\partial x_2}) = 2x_1'H_{12} + 2x_2'H_{22} \quad , \quad \frac{\partial x_1}{\partial x_2} = 0.$$

H_{22} is symmetric, these equations can be solved in closed form to give

$$x_1 = c_1 \tag{3.8}$$

$$w_0 = 2c_1'H_{12}x_2 + x_2'H_{22}x_2 + c_0 \tag{3.9}$$

e c_0 and c_1 are free parameters. But now, following standard procedure, we
ce that c_0 and c_1 can be expressed using (3.8) and (3.9) as functions of x_1
x_2, and that these functions form a set of mutually independent solutions to the
tions (3.6) and (3.7). Altogether, we now have as our desired transformation

$$z_0 = w_0 - 2x_1'H_{12}x_2 - x_2'H_{22}x_2 \tag{3.10}$$

$$z_1 = x_1 \tag{3.11}$$

$$z_2 = x_2 \tag{3.12}$$

e (3.10) and (3.11) follow from (3.8) and (3.9), and z_2 is chosen arbitrarily.
transformation is nonsingular as the Jacobian determinant equals unity.
It is important to note that the closed form solution (3.9) exists only if H_{22}

is symmetric. In the scalar control problem considered by Kelley, u_2 and x_2 are scalars, so that H_{22} is a scalar, and no difficulty arises. However, in general, there is no a priori symmetry constraint on H_{22} in (1.1) as it stands. Instead of (3.10), we consider the transformation

$$z_0 = w_0 - 2x_1'H_{12}x_2 - x_2'H_{22}^S x_2 \qquad (3.13)$$

where H_{22}^S is the symmetric part of H_{22}, and in the subsequent discussion and Appendix B, we show that the nonnegativity requirement (1.3) forces H_{22} to be symmetric. (An alternative approach to proving the symmetry is presented in Section 5). The symmetry property is actually the relevant generalized Legendre-Clebsch condition corresponding to (1.8) for $q = 1$.

From (3.11) and (3.12), the dynamics of z_1 and z_2 are seen to be identical to those of x_1 and x_2. Calculation of the dynamics of z_0 is straightforward. Identifying z_1 and z_2 with x_1 and x_2, we obtain

$$\dot{z}_0 = x_1'\hat{Q}x_1 + 2x_1'\hat{H}_1 x_2 + 2x_1\hat{H}_2 u_1$$

$$+ x_2'\hat{R}_1 x_2 + 2x_2'\hat{R}_2 u_1 + u_1'u_1 + x_2'H_{22}^A u_2 \qquad (3.14)$$

where

$$
\begin{aligned}
\hat{Q} &= Q_{11} - H_{12}F_{21} - F_{21}'H_{12}' \\
\hat{H}_1 &= Q_{12} - F_{11}H_{12} - H_{12}F_{22} - \dot{H}_{12} - F_{21}H_{22}^S \\
\hat{H}_2 &= H_{11} - H_{12}G_{21} \\
\hat{R}_1 &= Q_{22} - F_{12}H_{12} - H_{12}'F_{12} - F_{22}H_{22}^S - H_{22}^S F_{22} - \dot{H}_{22}^S \\
\hat{R}_2 &= H_{21} - H_{12}'G_{11} - H_{22}^S G_{21}
\end{aligned}
\qquad (3.15)
$$

H_{22}^A is the anti-symmetric part of H_{22}.

It is now clear from (3.14) that $\dot{z}_0$ is independent of the singular controls u_2, except for the final term $x_2'H_{22}^A u_2$.

This far we have considered only the dynamics of the new variables; it now remains to discuss the boundary conditions. We demand that the transformation (3.13) holds on the closed interval $[t_0, t_f]$, and we thus have at t_0 and t_f respectively

$$
\begin{aligned}
z_0(t_0) &= w_0(t_0) - 2x_1'(t_0)H_{12}(t_0)x_2(t_0) - x_2'(t_0)H_{22}^S(t_0)x_2(t_0) \\
z_0(t_f) &= w_0(t_f) - 2x_1'(t_f)H_{12}(t_f)x_2(t_f) - x_2'(t_f)H_{22}^S(t_f)x_2(t_f).
\end{aligned}
\qquad (3.16)
$$

Noting that $V[0, t_0, u(\cdot)] = x'(t_f)Sx(t_f) + w_0(t_f) - w_0(t_0)$ and that $x_1(t_0) = 0$ and $x_2(t_0) = 0$, we obtain from (3.14) and (3.16)

$$V[0, t_0, u(\cdot)] = x^{'}(t_f)Sx(t_f) + 2x_1^{'}(t_f)H_{12}(t_f)x_2(t_f) + x_2^{'}(t_f)H_{22}^{S}(t_f)x_2(t_f)$$

$$+ \int_{t_0}^{t_f}\{x_1^{'}\hat{Q}x_1 + 2x_1^{'}\hat{H}_1x_2 + 2x_1^{'}\hat{H}_2u_1 + x_2^{'}\hat{R}_1x_2$$

$$+ 2x_2^{'}\hat{R}_2u_1 + u_1^{'}u_1\}dt$$

$$+ \int_{t_0}^{t_f} x_2^{'}H_{22}^{A}u_2 dt \ . \tag{3.17}$$

In Appendix B, Lemma B.1, we show that the nonnegativity requirement (1.3) on $V[0, t_0, u(\cdot)]$ implies $H_{22}^{A} \equiv 0$ on the interval $[t_0, t_f)$. For continuous H_{22}^{A} this equality can then be extended to the closed interval $[t_0, t_f]$.

Introduce the notation

$$\hat{x} = x_1 \quad , \quad \hat{u} = [\hat{u}_1^{'} \quad \hat{u}_2^{'}] = [x_2^{'} \quad u_1^{'}]^{'},$$

$$\hat{F} = F_{11} \quad , \quad \hat{G} = [F_{12} \quad G_{11}] \quad , \quad \hat{H} = [\hat{H}_1 \quad \hat{H}_2] \quad , \tag{3.18}$$

$$\hat{R} = \begin{bmatrix} \hat{R}_1 & \hat{R}_2 \\ \hat{R}_2^{'} & I \end{bmatrix}.$$

Now, (3.17) can be written as

$$V[u(\cdot)] = [\hat{x}^{'}S_{11}\hat{x} + 2\hat{x}^{'}(S_{12} + H_{12})\hat{u}_1 + \hat{u}_1^{'}(S_{22} + H_{22})\hat{u}_1]_{t=t_f}$$

$$+ \int_{t_0}^{t_f}\{\hat{x}^{'}\hat{Q}\hat{x} + 2\hat{x}^{'}\hat{H}\hat{u} + \hat{u}^{'}\hat{R}\hat{u}\}dt \tag{3.19}$$

and (3.2) becomes

$$\dot{\hat{x}} = \hat{F}\hat{x} + \hat{G}\hat{u} \ . \tag{3.20}$$

Now we would like to replace the problem of finding necessary and sufficient conditions for (1.3) to hold, given (1.1) and (1.2) with $x(t_0) = 0$, by a problem of identical structure, save that the hat quantities only are involved. It is important to realize why we should want to do this. The state variable $\hat{x}$ evidently has lower dimension than the state variable x. Hence repetition of the cycle of reduction-to-standard form (with possible reduction of control space dimention) followed by state-space-dimension reduction via-Kelley-transformation must terminate, in one of three possible ways: either a zero dimension control variable is encountered, or a zero dimension state variable, or a nonsingular problem. In either of these three cases, helpful necessary and sufficient conditions for (1.3) follow.

Let us then return to an examination of the replacement hat problem. For the replacement problem to be of identical structure, in particular $\hat{u}$ is required to be a piecewise continuous control, and therefore u_2 constructed from (3.3) could contain delta functions at the discontinuities of u_1, i.e., x_2. Thus, the set of admissible control functions U for the original problem needs to be extended to contain delta functions if the attainable performance indices are to be the same. This is not a problem however; as a delta function can be constructed as a limit of continuous functions, the original nonnegativity requirement (1.3) is equivalent to

$$V[0, t_0, u(\cdot)] \geq 0 \qquad \text{for each} \quad u(\cdot) \in U^{\checkmark} \tag{3.21}$$

where $U^{\checkmark}$ is a suitably extended set of admissible controls $u(\cdot)$.

From (3.19) it is clear that the nonnegativity requirement (3.21) implies that the end point term of (3.19) must be bounded below for each $\hat{x}$, independently of $\hat{u}_1$. Necessary and sufficient conditions for this are that

i) $\quad S_{22} + H_{22}(t_f) \geq 0$ $\hfill$ (3.22)

ii) $\quad N[S_{22} + H_{22}(t_f)] \subseteq N[S_{12} + H_{12}(t_f)]$ $\hfill$ (3.23)

where N denotes null space. Now using a completion of the square type argument, the end point term in (3.19) can be written as

$$[\hat{u}_1 + (S_{22}+H_{22})^{\#}(S_{12}+H_{12})\hat{x}]^{\checkmark}(S_{22}+H_{22})[\hat{u}_1 + (S_{22}+H_{22})^{\#}(S_{12}+H_{12})\hat{x}]_{t=t_f}$$
$$+ \hat{x}^{\checkmark}[S_{11} - (S_{12}+H_{12})(S_{22}+H_{22})^{\#}(S_{12}+H_{12})^{\checkmark}]\hat{x}\Big|_{t=t_f} \tag{3.24}$$

where $\#$ denotes pseudo-inverse. With the notation

$$\hat{S} = S_{11} - [S_{12} + H_{12}(t_f)][S_{22} + H_{22}(t_f)]^{\#}[S_{12} + H_{12}(t_f)]^{\checkmark}$$

$$\hat{K} = -[S_{22} + H_{22}(t_f)]^{\#}[S_{12} + H_{12}(t_f)]^{\checkmark}$$

(3.19) becomes

$$V[0, t_0, u(\cdot)] = [\hat{u}_1 - \hat{K}\hat{x}]^{\checkmark}(S_{22} + H_{22})[\hat{u}_1 - \hat{K}\hat{x}]\Big|_{t=t_f} + \hat{x}^{\checkmark}\hat{S}\hat{x}\Big|_{t=t_f}$$
$$+ \int_{t_0}^{t_f}\{\hat{x}^{\checkmark}\hat{Q}\hat{x} + 2\hat{x}^{\checkmark}\hat{H}\hat{u} + \hat{u}^{\checkmark}\hat{R}\hat{u}\}dt \ . \tag{3.25}$$

Because we allow piecewise continuous controls $\hat{u}$ and because $u_1(t_f)$ appears in the end point term of (3.25) and has no effect on the value of the integral in (3.25), the minimization of (3.25) is carried out by separately minimizing the end point term involving $\hat{u}_1(t_f)$ and the remaining integral-plus-terminal-cost term. We can now state the following theorem which summarizes what we have to this point.

Theorem III.3.1: Assume continuity of F, G, Q and R, continuous differentiability of H and that the problem (1.1) through (1.3) with $x(t_0) = 0$ is in standard form. Further, with quantities $\hat{x}$, $\hat{u}$, $\hat{F}$, $\hat{G}$, $\hat{Q}$, $\hat{H}$, $\hat{R}$, $\hat{S}$ as defined earlier, set

$$\hat{V}[0, t_0, \hat{u}(\cdot)] = \hat{x}^{\checkmark}(t_f)\hat{S}\hat{x}(t_f) + \int_{t_0}^{t_f}\{\hat{x}^{\checkmark}\hat{Q}\hat{x} + 2\hat{x}^{\checkmark}\hat{H}\hat{u} + \hat{u}^{\checkmark}\hat{R}\hat{u}\}dt. \tag{3.26}$$

Then

$$V[0, t_0, u(\cdot)] \geq 0 \quad \text{for each} \quad u \in U, \quad \text{subject to} \quad (1.2) \text{ with} \quad x(t_0) = 0$$

if and only if

(a) $\hat{V}[0, t_0, \hat{u}(\cdot)] \geq 0$ for each $\hat{u} \in U$, subject to (3.20) with $\hat{x}(t_0) = 0$

(b) $H_{22}(t)$ is symmetric for each $t \in [t_0, t_f]$

(c) $S_{22} + H_{22}(t_f) \geq 0$

(d) $N[S_{22} + H_{22}(t_f)] \subseteq N[S_{12} + H_{12}(t_f)].$

arks 3.1: 1. Recall from the last section that to put the given problem into ndard form, it is necessary to make some assumptions on the ranks and different-ility of matrices constructible from the coefficient matrices F. G. etc.

2. This theorem is independent of any controllability assumption such (II.3.5). However, for the reduction from nonstandard to standard form as set in Section 2, (II.3.5) guarantees the retention of the controllability property thus precludes the possibility of G and H being identically zero in the stand-form.

3. A necessary condition for $\hat{V}[0, t_0, \hat{u}(\cdot)] \geq 0$ for each $\hat{u} \in U$ sub-: to (3.20) with $\hat{x}(t_0) = 0$ is $\hat{R}(t) \geq 0$ on $[t_0, t_f]$, the classical Legendre-sch condition. For the original problem (without the hat superscripts) this becomes generalized Legendre-Clebsch condition corresponding to (1.5).

Theorem 3.1 says that our original singular problem (1.3) is equivalent to an tical, though possibly nonsingular, problem of lower state dimension (condition of Theorem 3.1) plus side conditions ((b), (c) and (d) of Theorem 3.1). If R ingular on the interval $[t_0, t_f]$ and the various differentiability and rank mptions hold, the process of conversion to standard form with possible elimination some controls, followed by application of Theorem 3.1 can be repeated to produce yet wer dimensional problem and further side conditions. Now, since the state dimension lowered at each application of Theorem 3.1, the process must end when either the te dimension shrinks to zero, or the problem becomes nonsingular, or G and H me zero in standard form. However, should the controllability assumption (II.3.5) n force, this third possibility cannot occur [see Remark 3.1.2 above].

In case the state dimension shrinks to zero, necessary and sufficient conditions trivial; in case a nonsingular problem is obtained, necessary and sufficient itions are given by the classical Jacobi conjugate point condition in the form of ccati equation having no escape times on the interval $(t_0, t_f]$ – see Cor. II.3.2. lly for G and H zero in standard form a necessary and sufficient condition is nonnegativity of $R(t)$ on $[t_0, t_f]$.

In Section 4, Theorem 3.1 is extended to the general linear-quadratic control lem with no end-point constraints. Computation of optimal controls and the corresponding optimal cost is also discussed.

4. COMPUTATION OF OPTIMAL CONTROL AND PERFORMANCE INDEX

In this section we are interested in the general linear-quadratic control probl
which can be stated as:

> Find necessary and sufficient conditions for (1.1) to be bounded
> below independently of $u \in U$, subject to (1.2) with initial
> condition $x(t_0) = x_0$ not necessarily zero. Moreover, when the $\qquad$ (4.1)
> lower bound exists, find a minimizing (optimal) control and the
> corresponding minimal (optimal) cost.

Here we shall solve this problem by extending the results of the previous section.
In Section 5 we will again derive the optimal control and the optimal cost using
Theorem II.3.3 of the previous chapter and the Anderson-Moylan algorithm.

As in Section 2, we assume that (1.1) and (1.2) are in standard form, that the
various differentiability assumptions hold and that the state and performance index
transformation is described by equations (3.10)-(3.12). Recalling that $x(t_0) = x_0$
is now arbitrary but fixed, we obtain the additional <u>fixed</u> term (previously zero).

$$-2x_1'(t_0)H_{12}(t_0)x_2(t_0) - x_2'(t_0)H_{22}^S(t_0)x_2(t_0) \qquad (4.2)$$

in the computation of $V[x_0, t_0, u(\cdot)]$ in (3.17). Now if the infimum of (1.1) subject to (1.2) is finite for all $x(t_0)$, this implies $V^*[0, t_0] = 0$ and that
$V[0, t_0, u(\cdot)] \geq 0$ for all $u(\cdot)$. As argued in the last section, it follows $H_{22}(t_.$
is symmetric on $[t_0, t_f]$.

Arguing further as in the previous section, we have the following extension to
Theorem 3.1.

> <u>Theorem III.4.1</u>: With the same notation and assumptions as for Theorem 3.1 but
> allowing free $x(t_0)$ (and again noting that controllability is not required),
> we obtain
>
> $V[x_0, t_0, u(\cdot)]$ subject to (1.2) is bounded below for each fixed $x(t_0)$,
> independently of $u \in U$
>
> if and only if
>
> (a) $\hat{V}[\hat{x}_0, t_0, \hat{u}(\cdot)]$ subject to (3.20) is bounded below for each fixed $\hat{x}_0$,
> independently of $\hat{u} \in U$
>
> (b) $H_{22}(t)$ is symmetric for each $t \in [t_0, t_f]$
>
> (c) $S_{22} + H_{22}(t_f) \geq 0$

(d) $N[S_{22} + H_{22}(t_f)] \subseteq N[S_{12} + H_{12}(t_f)]$.

extending the discussion of Section 3, we perform a series of such transformat-
and applications of Theorem 4.1 in conjunction with the transformation of the
icient matrices to standard form until we obtain either a problem of zero state
sion, or a nonsingular problem or one with G and H being zero. For the first
ssibilities, necessary and sufficient conditions for (4.1) are known. For the
possibility, which would be ruled out by a controllability assumption (II.3.5),
essary and sufficient condition is that $R(t) \geq 0$ on $[t_0, t_f]$. The minimum
would then be $x'(t_0)\{\Phi'(t_f, t_0)S\Phi(t_f, t_0) + \int_{t_0}^{t_f}\Phi'(\tau, t_0)Q(\tau)\Phi(\tau, t_0)d\tau\}x(t_0)$.
Now, to calculate the minimizing control and the corresponding minimal cost, we
backwards from either the nonsingular, zero state dimension or zero input dimension
em, minimizing at each successive stage. For the purpose of illustration, suppose
that after one transformation the problem is nonsingular, i.e. $\hat{R}(t) > 0$ on
$_f]$. Then the necessary and sufficient condition for $\hat{V}^*[\hat{x}_0, t_0]$ to be finite is
the Riccati equation

$$-\dot{\hat{P}} = \hat{P}\hat{F} + \hat{F}'\hat{P} + \hat{Q} - (\hat{P}\hat{G}+\hat{H})\hat{R}^{-1}(\hat{P}\hat{G}+\hat{H})' \quad , \quad \hat{P}(t_f) = \hat{S} \tag{4.3}$$

$\hat{P}$ is a symmetric square matrix of appropriate dimension, has no escape times
interval $[t_0, t_f]$. From standard linear regulator theory we know that the
l control for the cost term $\hat{V}[\hat{x}_0, t , \hat{u}(\cdot)]$ subject to (3.19) with $\hat{x}(t_0) = \hat{x}_0$
ecessarily zero, is

$$\hat{u}^*(t) = \hat{L}(t)\hat{x}(t) \quad \text{for} \quad t \in [t_0, t_f] \tag{4.4}$$

$\hat{L} = -\hat{R}^{-1}(\hat{G}'\hat{P} + \hat{H}')$ and the corresponding minimum cost is

$$\hat{V}[\hat{x}_0, t_0, \hat{u}^*(\cdot)] = \hat{x}'(t_0)\hat{P}(t_0)\hat{x}(t_0). \tag{4.5}$$

er, we also need to separately minimize a terminal point term occurring in
$t_0, u(\cdot)]$ but not in $\hat{V}[\hat{x}_0, t_0, \hat{u}(\cdot)]$; this terminal point term is shown in
for the case when $x_0 = 0$, but clearly takes the same form independently of
lue of x_0. The separate minimization gives the optimal value for the control
$t = t_f$,

$$\hat{u}_1^*(t_f) = \hat{K}\hat{x}(t_f) \tag{4.6}$$

e corresponding minimal cost for the terminal point term of zero [see (3.25)].
timal value for $\hat{u}_2(t_f)$ is seen to be indeterminate, the most convenient value
that defined by (4.4). Now considering the optimal control at t_0 we see that
is specified as $x_2(t_0)$. Again, we also have $\hat{u}_2^*(t_0)$ arbitrary, the most
ient value being that defined by (4.4).
Je now combine the optimal cost and control from the separate optimization prob-
o obtain the optimal cost and control, in terms of the hat quantities, for the

problem (4.1). From (4.5) and (4.2), we can write the optimal value for $V[x_0, t_0, u(\cdot)]$ as

$$[x_1^*(t_0) \quad x_2^*(t_0)] \begin{bmatrix} \hat{P}(t_0) & -H_{12}(t_0) \\ -H_{12}^*(t_0) & -H_{22}(t_0) \end{bmatrix} \begin{bmatrix} x_1(t_0) \\ x_2(t_0) \end{bmatrix} \qquad (4.7)$$

while the optimal control constructed from (4.4) and (4.6) is

$$\hat{u}^*(t) = \hat{L}(t)\hat{x}(t) \qquad \text{for} \quad t \in (t_0, t_f)$$

$$\hat{u}_1^*(t_0) = x_2(t_0)$$

$$\hat{u}_1^*(t_f) = \hat{K}\hat{x}(t_f) \qquad (4.8)$$

$$\hat{u}_2^*(t_0) \quad \text{and} \quad u_2^*(t_f) \quad \text{determined as discussed above.}$$

The computation of the optimal control for the problem (4.1) is then completed by using (3.3) to determine u_2^* from x_1^*, x_2^* and u_1^* (the last three quantities being in hat notation, $\hat{x}^*$, $\hat{u}_1^*$ and $\hat{u}_2^*$); the u_1^* part of the control u^* is already determined by $\hat{u}_2^*$. The possible occurrence of delta functions in the optimal control u^* at both the initial and final points of the interval $[t_0, t_f]$ is now apparent since from (4.8) there is the possibility of jumps in the optimal control $\hat{u}^*$ at the end points of $[t_0, t_f]$. To prevent the possibility of delta functions occurring within the interval (t_0, t_f), we demand that $\hat{L}(t)$, which is constructed from $\hat{P}$, $\hat{R}$ and $\hat{G}$, be continuously differentiable throughout the interval. Finally, the arbitrary nature of $\hat{u}_2^*(t_0)$ and $\hat{u}_2^*(t_f)$ introduces nonuniqueness into the choice of optimal control, in the form of nonuniqueness in the delta functions at t_0 and t_f.

Above, we have discussed the procedure applying when the transformed problem is nonsingular. Suppose now that the transformed problem has zero state dimension. Then $\hat{V}[\hat{x}_0, t_0, u(\cdot)]$ is just $\int_{t_0}^{t_f} \hat{u}^*\hat{R}\hat{u} \, dt$; a necessary and sufficient condition for this to be bounded below is that $\hat{R} \geq 0$ on $[t_0, t_f]$. For $\hat{R} > 0$, the optimal (and clearly unique) control is $\hat{u}^*(t) \equiv 0$. However, for $\hat{R}$ of rank s along $[t_0, t_f]$, a transformation to standard form makes it clear that only the first s components of u are required to be set to zero, the remaining components of u being arbitrary. Calculation of the optimal control and cost can now be carried out along the lines of the procedure discussed for the case of the transformed problem being nonsingular.

Finally for G and H being zero in standard form, the minimum value was described earlier. The corresponding optimal control is then calculated as in the previous paragraphs.

Any of the three cases discussed above could arise as the first step in the backward procedure required to calculate the optimal control and cost for a problem where more than one transformation is needed to obtain a nonsingular, zero state dimension, or zero input dimension problem. To complete the discussion we therefore need to look

fly at the procedure for calculating the optimal control and cost for a singular
lem from the optimal quantities for a singular problem of lower (but nonzero)
e dimension, as set out in Theorem 4.1. The optimal control for the lower dimen-
al problem is assumed known, and is continuously differentiable on (t_0, t_f)
the possibility of delta functions and derivatives of delta functions at the end
ts, and the further possibility of jumps at the end points due to the minimization
edure at the stage under discussion. Noting (3.3), one sees that the optimal
rol for the higher dimensional problem will now contain derivatives of those delta
tions and jumps in the optimal control for the lower dimensional problem. The
ulation of the optimal cost would proceed along the lines of (4.7) with blocks
he matrix defining the quadratic performance index being uniquely identified by
end point conditions.

The optimal control is not necessarily unique though certain components of it
such as the control derived from the nonsingular problem and the optimal value
the end point. However, nonuniqueness can arise in various ways, the main ones
g: reduction of control dimension in bringing the problem to standard form,
inating with singular problem with zero state dimension, and certain end point
rols not appearing in the coast. Note also, that as shown by looking at the
mal cost (4.7), part of the performance matrix is determined uniquely by parts of
), while the remaining part is determined uniquely by the Riccati equation (4.3).
he next section, this will appear in the derivation more naturally than in the
e.

Finally, in most discussions of problems in singular control there arises the
tion of the definition of singular strips, i.e. subspaces on which the state
or is concentrated when the control is optimal. In the derivation we have presented
contrast to that in [19]), the singular strip is closely related to the subspace
he original state space described by the states in the terminating problem, whether
e a nonsingular problem of nonzero state dimension or a zero state dimension prob-
In the former case, the definition of the singular strip in terms of the co-
nates describing the original state space can be quite complicated since the
ous transformations performed in arriving at the terminating problem must be
ied in reverse order. However, the latter case is simple; the singular strip is
the origin, the unique zero dimensional subspace of the original state space.
s also easy to interpret the occurrence of delta functions and their derivatives
he end points of the optimal control. They allow the instantaneous transfer of
initial state onto the singular strip at the initial time t_0 and a jump off the
ular strip at the final time t_f.

SOLUTION VIA RIEMANN-STIELTJES INEQUALITY

In the solution of the linear-quadratic control problem presented in Section 4, the reduction in state dimension and the calculation of the optimal control appear in a direct manner, with the computation of the optimal cost completing the solution of problem. Here we present an alternative derivation of the results of Section 4 employing the Anderson-Moylan algorithm in conjunction with Theorem II.3.3. By this method manipulations are made on the matrix measure involving P in order to compute a matrix solution of the integral matrix inequality. In contrast to the method presented in Section 4, the state transformation and optimal control are not part of the main algorithm.

Recall that Theorem II.3.3 connnects the linear-quadratic problem (4.1) and necessary and sufficient conditions involving the Riemann-Stieltjes inequality. In general, there can be many matrices $P(\cdot)$ satisfying the Riemann-Stieltjes inequality but as we have already shown in Theorem II.5.1, there is a maximal solution and this defines the performance index for the associated control problem.

For convenience, we rewrite the Riemann-Stieltjes inequality, namely,

$$\int_{t_1}^{t_2} [v' \quad u'] \begin{bmatrix} dP + (PF + F'P + Q)dt & (PG + H)dt \\ (PG + H)'dt & R\,dt \end{bmatrix} \begin{bmatrix} v \\ u \end{bmatrix} \geq 0 \qquad (5.1)$$

for all continuous $v(\cdot)$, for all piecewise continuous $u(\cdot)$ and for all $[t_1, t_2]$ $\subset [t_0, t_f]$. In addition the end point condition

$$P(t_f) \leq S \qquad (5.2)$$

must be satisfied.

As in the previous sections, assume that G and R are given in standard form and that the corresponding partitioning of the various matrices and vectors hold. Substituting into (5.1), defining $w' = [v_1' \quad v_2' \quad u_1']$ and with the obvious definition of dY, we obtain

$$\int_{t_1}^{t_2} w'dYw + 2\int_{t_1}^{t_2} v_1'(P_{12} + H_{12})u_2dt + 2\int_{t_1}^{t_2} x_2'(P_{22} + H_{22})u_2dt \geq 0 \qquad (5.3)$$

from which we are able to conclude that

$$P_{12}(t) + H_{12}(t) = 0 \qquad \text{on} \qquad (t_0, t_f) \qquad (5.4)$$

$$P_{22}(t) + H_{22}(t) = 0 \qquad \text{on} \qquad (t_0, t_f). \qquad (5.5)$$

To see that (5.5), holds, suppose that there exist $[t_1, t_2] \subset [t_0, t_f]$, $v_2(\cdot)$ and $u_2(\cdot)$ such that

$$\int_{t_1}^{t_2} v_2'(P_{22} + H_{22})u_2dt \qquad (5.6)$$

is not zero. Then choose $v_1(t) \equiv 0$ on $[t_1, t_2]$, so that the middle term in the inequality (5.3) is zero for all $u_2(\cdot)$. Since the first term of (5.3) is unaffected by $u_2(\cdot)$, it is then clear that for suitable scaling of $u_2(\cdot)$ we obtain a contration to the inequality (5.3). Thus, (5.6) must be zero for any $[t_1, t_2] \subset [t_0, t_f]$, any continuous $v_2(\cdot)$ and for any piecewise continuous $u_2(\cdot)$. However, $P_{22}(\cdot)$ of bounded variation and therefore is continuous except at a countable number of nts in the interval $[t_0, t_f]$. Thus, with $H_{22}(t)$ continuous, we conclude that 5) holds at all points of continuity of $P_{22}(t)$ in the interval (t_0, t_f).

We can extend the validity of (5.5) to the entire interval (t_0, t_f) in the lowing way. Suppose that t_2 is a point of discontinuity of $P(\cdot)$. Because $P(\cdot)$ monotone, it has left and right limits at t_2; as noted in Chapter II, see Lemma 3.5, jumps in $P(\cdot)$ must be nonnegative matrices. Therefore jumps in $P_{22}(\cdot)$ must ewise be nonnegative, i.e. $\lim_{t \uparrow t_2} P_{22}(t) \le \lim_{t \downarrow t_2} P_{22}(t)$. However, both these limits t be $H_{22}(t_2)$, in view of the continuity of $H_{22}(\cdot)$ and the almost everywhere ality of P_{22} and H_{22}. Thus $P_{22}(t_2) = H_{22}(t_2)$, with $P_{22}(\cdot)$ possessing no ps. This concludes the proof of (5.5). In a similar manner, we can argue that 4) must also hold.

We have now identified the blocks P_{12} and P_{22} of P uniquely for any P satis- ng (5.1); any nonuniqueness can only occur in the P_{11} block. Moreover, P_0 is metric on (t_0, t_f), implying by (5.5) the symmetry of H_{22} as a necessary con- ion for the solvability of the optimal control problem.

We can also show that the equalities (5.4) and (5.5) extend to the point t_0 in $P(t) = P^*(t)$, where $P^*(\cdot)$ is as defined previously. Recall that the Riemann- ltjes integral inequality also implies - see Lemma II.3.5 - that all jumps in $P(\cdot)$ satisfying the inequality must be nonnegative, i.e. $P(t_-) \le P(t) \le P(t_+)$ so

$$
\begin{bmatrix} P_{11}(t_{0_+}) & -H_{12}(t_0) \\ -H_{12}'(t_0) & -H_{22}(t_0) \end{bmatrix} \ge \begin{bmatrix} P_{11}(t_0) & P_{12}(t_0) \\ P_{12}'(t_0) & P_{22}(t_0) \end{bmatrix} .
$$

ts clear that taking $P_{12}(t_0) = -H_{12}(t_0)$, $P_{22}(t_0) = -H_{22}(t_0)$ is consistent with Riemann-Stieltjes integral equality, and by the maximal property of $P^*(\cdot)$, and articular $P^*(t_0)$, we must then have $P_{12}^*(t_0) = -H_{12}(t_0)$, $P_{22}^*(t_0) = -H_{22}(t_0)$.

A study of the right hand end-point t_f leads to

$$
S = \begin{bmatrix} P_{11}^*(t_f) & P_{12}^*(t_f) \\ P_{12}^*(t_f) & P_{22}^*(t_f) \end{bmatrix} \ge \begin{bmatrix} P_{11}(t_f) & P_{12}(t_f) \\ P_{12}'(t_f) & P_{22}(t_f) \end{bmatrix} \ge \begin{bmatrix} P_{11}(t_f-) & P_{12}(t_f-) \\ P_{12}'(t_f-) & P_{22}(t_f-) \end{bmatrix}
$$

$$
= \begin{bmatrix} P_{11}(t_f-) & -H_{12}(t_f) \\ -H_{12}'(t_f) & -H_{22}(t_f) \end{bmatrix}
$$

hat $N[S_{22} + H_{22}(t_f) \subseteq [S_{12} + H_{12}(t_f)]$ and $S_{22} + H_{22}(t_f) \ge 0$. Further, returning to (5.3) and the definition of dY we have

$$dY = \begin{bmatrix} dY_{11} & dY_{12} & dY_{13} \\ dY_{12}' & dY_{22} & dY_{23} \\ dY_{13}' & dY_{23}' & I \end{bmatrix} \qquad (5.7)$$

where

$$dY_{11} = dP_{11} + (Q_{11} + P_{11}F_{11} + F_{11}'P_{11} + P_{12}F_{21} + F_{21}'P_{12}')dt$$

$$dY_{12} = dP_{12} + (Q_{12} + P_{11}F_{12} + P_{12}F_{22} + F_{11}'P_{12} + F_{21}'P_{22})dt$$

$$dY_{22} = dP_{22} + (Q_{22} + P_{22}F_{22} + F_{22}'P_{22} + P_{12}'F_{12} + F_{12}'P_{12})dt$$

$$dY_{13} = (P_{11}G_{11} + P_{12}G_{21} + H_{11})dt$$

$$dY_{23} = (P_{12}'G_{11} + P_{22}G_{21} + H_{21})dt \ . \qquad (5.8)$$

Now, assuming the differentiability of H, we combine the definitions (3.18) with the above to obtain

$$dY = \begin{bmatrix} d\hat{P} + (\hat{P}\hat{F} + \hat{F}'\hat{P} + \hat{Q})dt & (\hat{P}\hat{G} + \hat{H})dt \\ (\hat{P}\hat{G} + \hat{H})'dt & \hat{R}dt \end{bmatrix} \qquad (5.9)$$

where $\hat{P} = P_{11}$.

Observing that the Riemann-Stieltjes integral of (5.9) has the same form as the original integral (5.1), we attempt to find the relevant minimization problem [of the same form as (4.1)] corresponding to (5.9). However, given our development of Sections 3 and 4 it is clear that with the transformation (3.10)-(3.12), the definitions of $\hat{x}$ and $\hat{u}$ as in (3.18) and $\hat{S}$ as defined in Section 3, the minimization problem is just that described in part (a) of Theorem 4.1.

The above discussion leads us to

Theorem III.5.1: With the same assumptions as in Theorem 3.1, there exists an $n \times n$ matrix $P(t)$, symmetric and of bounded variation of $[t_0, t_f]$ such that with arbitrary $[t_1, t_2] \subset [t_0, t_f]$, (5.1) and (5.2) hold if and only if there exists a matrix $P(t)$ of appropriate dimension, symmetric and of bounded variation on $[t_0, t_f]$ such that

(a) $\hat{P}(t_f) \le \hat{S}$ \hfill (5.10)

(b) $\displaystyle\int_{t_1}^{t_2} [\hat{v}' \quad \hat{u}'] \begin{bmatrix} d\hat{P} + (\hat{P}\hat{F} + \hat{F}'\hat{P} + \hat{Q})dt & (\hat{P}\hat{G} + \hat{H})dt \\ (\hat{P}\hat{G} + \hat{H})'dt & \hat{R}dt \end{bmatrix} \begin{bmatrix} \hat{v} \\ \hat{u} \end{bmatrix} \ge 0$ \hfill (5.11)

for all continuous $\hat{v}(\cdot)$ and for all piecewise continuous $\hat{u}(\cdot)$

(c) $H_{22}(t)$ is symmetric on $[t_0, t_f]$

(d) $S_{22} + H_{22}(t_f) \ge 0$

(e) $N[S_{22} + H_{22}(t_f)] \subseteq N[S_{12} + H_{12}(t_f)]$.

Again, as in the previous sections, the application of Theorem 5.1 and the trans-
-mation to standard form may need to be made a number of times, terminating with
her a zero dimensional $\hat{P}$ in which case the original P would be completely and
.quely identified by a series of equalities such as (5.4) and (5.5), or a problem
h $\hat{P}$ of positive dimension with $\hat{R}$ nonsingular, or in transforming from nonstand-
to standard form a problem with G and H zero may arise. For the second case,
we know, (5.10) and (5.11) have a solution $\hat{P}$ if and only if the Riccati equation
3) has no escape times on $[t_0, t_f]$. Moreover, the unique solution $\hat{P}$ of the
:cati equation is the maximal of many possible solutions of (5.10) and (5.11) – see
orem II.5.1. Finally, $\hat{P}$ as calculated from the Riccati equation is connected to
optimal cost via the standard quadratic form. Tracing back to the original control
blem, the solution P of (5.1) so generated defines the optimal cost for each
$_0$) for problem (4.1).

For the third case when G and H are zero, we have noted earlier what $\hat{P}$ is.
in, one can trace back to a solution P of the original control problem.

SUMMARIZING REMARKS

In this chapter, we have given an algorithmic procedure for computing a matrix,
existence of which is guaranteed by the nonnegativity of a certain functional.
irectly, this gives a procedure for checking the nonnegativity of the functional.
ond, we have shown how this algorithm can also be used in computing the optimal
formance index and optimal control (the latter possibly not being unique) for
ear-quadratic singular optimal control problems. Several key properties of the
orithm are: its capacity to handle vector control problems; its linkage with, on
one hand, other and possibly less complete approaches to the singular control
blem, and on the other hand, with the singular time-varying covariance factorizat-
problem; its illumination of singular strips; its disadvantage, viz., a require-
t that the ranks of certain matrices remain constant over the interval of interest,
that certain matrices enjoy differentiability properties.

There is another possible approach to the optimal control problem which we have
mentioned to this point. By a standard completion of the square device, one can
racterise the optimal control, if it exists, in open-loop form as the solution of
inear Fredholm integral equation which is only of the second kind in case the
imal control problem is nonsingular. A solution procedure for the dual singular
blem (arising in detection theory) is studied in [28] and could presumably be mod-
ed to deal with the control problem.

APPENDIX III.A

DOLEZAL'S THEOREM

The following statement of Dolezal's Theorem is drawn from [24, 25].

Theorem III.A.1 Let $A(\cdot)$ be an $r \times r$ matrix with entries possessing continuous p-th order derivatives in $[a, b]$, and with rank $A(t) = h$ for all $t \in [a, b]$. Then there exists an $r \times r$ matrix $M(\cdot)$ with entries possessing continuous p-th order derivatives in $[a, b]$ and with $M(t)$ nonsingular for all $t \in [a, b]$ such that $A(t)M(t) = [B(t) \vdots 0]$ for all $t \in [a, b]$. Here, $B(\cdot)$ is an $r \times h$ matrix with rank $B(t) = h$.

If $A(t) = A'(t)$ for all t and M is constructed as above, it is clear that $M'[B \vdots 0]$ must be symmetric, from which it follows that

$$M'(t)A(t)M(t) = \begin{bmatrix} C(t) & \vdots & 0 \\ - - - & + & - - - \\ 0 & \vdots & 0 \end{bmatrix}$$

with $C(t)$ of dimension $h \times h$ and nonsingular for all t.

If in addition $A(t)$ is nonnegative definite, $C(t)$ is positive definite. There exists a triangular $D(t)$ with entries expressible in terms of the entries of $C(t)$ and inheriting their differentiability property such that $C(t) = D'(t)D(t)$, [26]. Then with $N(t) = M(t)D^{-1}(t)$, possessing entries with continuous p-th order derivatives and nonsingular for all $t \in [a, b]$, one has

$$N'(t)A(t)N(t) = \begin{bmatrix} I & \vdots & 0 \\ - - & + & - - \\ 0 & \vdots & 0 \end{bmatrix}.$$

APPENDIX III.B

SYMMETRY CONDITION

The result proved here is an extension of a similar result needed for the dual covariance factorization problem [5].

With notation independent of that in the main portion of this chapter, define

$$V[\xi, u] = [2z_1'S_2z_2 + z_2'S_3z_2]_{t=t_f} + \int_{t_0}^{t_f} z_2'Au \, dt$$

$$+ \int_{t_0}^{t_f} \{z_1'Qz_1 + 2z_1'Hz_2 + z_2'Rz_2\}dt \tag{B.1}$$

with z_1, z_2 and u related by

$$\dot{z}_1 = F_{11}z_1 + F_{12}z_2 \qquad z_1(t_0) = \xi_1 \tag{B.2}$$

$$\dot{z}_2 = F_{21}z_1 + F_{22}z_2 + u \qquad\qquad z_2(t_0) = \xi_2. \qquad\qquad (B.3)$$

sume Q, H, R, F and A are continuous matrices on $[t_0, t_f]$, Q and R are metric and A is antisymmetric, and S_2, S_3 are constant. The dimensions of all ntities are arbitrary provided consistency is maintained.

<u>Lemma III.B.1</u>: If $A(t) \not\equiv 0$ on $[t_0, t_f]$, there exists a bounded piecewise continuous control $\bar{u}(\cdot)$ such that $V[0, \bar{u}] < 0$.

oof: Since A is antisymmetric, the lemma does not apply for scalar u. Also, if $\cdot) \equiv 0$ on $[t_0, t_f)$ then by continuity of $A(t)$ we have $A(t_f) \equiv 0$.

Therefore, consider $\sigma \in [t_0, t_f)$, k and j distinct indices such that $(\sigma) \neq 0$ where $a_{kj}(\sigma)$ is the k-j^{th} element of $A(\sigma)$. Without loss of general-, assume $a_{kj}(\sigma) > 0$.

Let the transition matrix associated with (B.2) and (B.3) be $\Phi(t, \tau)$. Then, any given $\varepsilon > 0$, there exists $\delta(\varepsilon) > 0$ such that for all t, τ satisfying t, $\tau \le \sigma + \delta$ we have $||\Phi(t, \tau) - I|| \le \varepsilon$ where $||\Phi|| = $ maximum of the norms the elements of Φ.

Let $\bar{\delta}$ be any positive number with $\bar{\delta} \le \delta(\varepsilon)$, and choose the control $\bar{u}(\cdot)$ to identically zero except that

$$\bar{u}_k(\tau) = -\omega\cos\omega(\tau-\sigma)$$
$$\qquad\qquad\text{for } \tau \in [\sigma, \sigma+\bar{\delta}]. \qquad\qquad (B.4)$$
$$\bar{u}_j(\tau) = -\omega\sin\omega(\tau-\sigma)$$

ng (B.2), (B.3) and (B.4) and taking $\xi = 0$ we can write

$$\begin{bmatrix} z_1(t) \\ \\ z_2(t) - \int_{t_0}^{t}\bar{u}(\tau)d\tau \end{bmatrix} = \int_{t_0}^{t} \{\Phi(t, \tau) - I\} \begin{bmatrix} 0 \\ \\ \bar{u}(\tau) \end{bmatrix} d\tau. \qquad (B.5)$$

m (B.5), we conclude that on the interval $[\sigma, \sigma+\bar{\delta}]$ we have

$$|z_{2k}(t) + \sin \omega(t-\sigma)| \qquad \le 2\varepsilon np$$
$$|z_{2j}(t) + 1 - \cos \omega(t-\sigma)| \le 2\varepsilon np$$
$$\qquad\qquad\qquad\qquad\qquad\qquad\qquad (B.6)$$
$$|z_{2\ell}(t)| \le 2\varepsilon np \qquad\qquad \ell \neq k, j$$
$$|z_{1r}(t)| \le 2\varepsilon np \quad , \quad 1 \le r \le \dim z_1$$

re

$$p = \dim z_1 + \dim z_2$$

$n = $ smallest integer greater than or equal to $\dfrac{2\omega\delta}{\pi}$.

Now consider z_1 and z_2 on the interval $[\sigma+\bar{\delta}, t_f]$. Set $z' = [z_1'\quad z_2']$ and let $K_1 = \max\limits_{t,\tau \ \varepsilon \ [t_0, \ t_f]} ||\Phi(t, \ \tau)||$; this quantity is well defined by the continuity of $\Phi(t, \ \tau)$. If we choose ω such that

$$\omega\bar{\delta} = 2\pi \tag{B.7}$$

then from (B.6) we see that $|z_i(\alpha+\bar{\delta})| \leq 8\varepsilon p$ for every component of the vector $z(\sigma+\bar{\alpha})$. Finally, since $z(t) = \Phi(t, \ \sigma+\bar{\delta})z(\sigma+\bar{\delta})$ on $[\sigma+\bar{\delta}, \ t_f]$ we have

$$|z_i(t)| \leq 8\varepsilon pK_1 \quad \text{on} \quad [\sigma+\bar{\delta}, \ t_f] \ . \tag{B.8}$$

Just as we derived the inequalities (B.6) from (B.5) we have

$$\left| \int_{t_0}^{t_f} z_2'A\bar{u} \ dt + z_{kj}(\sigma)[\omega\bar{\delta} - \sin\omega\bar{\delta}] \right| \leq K_2\varepsilon\omega\bar{\delta} \tag{B.9}$$

where K_2 is a constant. Some care needs to be taken with the bound in (B.9). In particular, $\bar{\delta}$ needs to be chosen so that $|a_{kj}(t) - a_{kj}(\sigma)| \leq \varepsilon$ for $\sigma \leq t \leq \sigma+\bar{\delta}$; this causes no problems since A is continuous and $\bar{\delta}$ is arbitrary. Setting $\omega\bar{\delta} = 2\pi$ (B.9) becomes

$$\left| \int_{t}^{t_f} z_2'A\bar{u} \ dt + 2\pi a_{kj}(\sigma) \right| \leq 2\pi\varepsilon K_2 . \tag{B.10}$$

From (B.9) the first term on the right side of (B.1) is upper bounded by $K_3\varepsilon^2$ for some constant K_3; from (B.10) the first integral on the right side of (B.1) is upper bounded by $-2\pi a_{kj}(\sigma) + 2\pi\varepsilon K_2$; and from (B.8) and (B.6) the final integral in (B.1) is upper bounded by $K_4\bar{\delta} + K_5\varepsilon^2$ for constants K_4 and K_5. Combining we have

$$V[0, \ \bar{u}] \leq K_3\varepsilon^2 - 2\pi a_{kj}(\sigma) + 2\pi\varepsilon K_2 + K_4\bar{\delta} + K_5\varepsilon^2 . \tag{B.11}$$

Because $a_{kj}(\sigma) > 0$ and $\bar{\delta}$ and ε are arbitrarily small, we see that $V[0, \ \bar{u}]$ can be guaranteed negative. This proves the lemma.

REFERENCES

[1] A.E. Bryson and Y.C. Ho., Applied Optimal Control, Ginn and Co., Mass., 1969.

[2] J.C. Willems, "Least squares stationary optimal control and the algebraic Riccati equation", IEEE Trans. Automatic Control, Vol. AC-16, December 1971, pp. 621-634.

[3] J.B. Moore and B.D.O. Anderson, "Extensions of quadratic minimization theory: I. Finite time results", Int. J. Control, Vol. 7, 1968, pp. 465-472.

[4] B.D.O. Anderson, J.B. Moore and S.G. Loo, "Spectral factorization of time-varying covariance functions", IEEE Trans. Information Theory, Vol. IT-15, 1969, pp. 550-557.

[5] B.D.O. Anderson and P.J. Moylan, "Spectral factorization of a finite-dimensional nonstationary matrix covariance", IEEE Trans. Automatic Control, Vol. AC-19, 1974, pp. 680-692.

N. Krasner, "Semi-separable kernels in linear estimation and control", *Report No. 7001-7*, Information Systems Laboratory, Stanford, Calif., May 1974.

B.D.O. Anderson and P.J. Moylan, "Synthesis of linear time-varying passive networks", *IEEE Trans. Circuits and Systems*, Vol. CAS-21, 1974, pp. 678-687.

D.H. Jacobson, "Totally singular quadratic minimization problems", *IEEE Trans. Automatic Control*, Vol. AC-16, 1971, pp. 651-658.

D.J. Bell, "Singular problems in optimal control - a survey", *Control Systems Centre Report, No. 249*, University of Manchester Institute of Science and Technology, England, 1974.

D.J. Bell and D.H. Jacobson, Singular Optimal Control Problems, Academic Press, New York, 1975.

H.M. Robbins, "A generalized Legendre-Clebsch condition for the singular cases of optimal control", *IBM J. Res. Develop.*, Vol. 3, 1967, pp. 372.

H.J. Kelley, "A second variation test for singular extremals", *AIAA J.*, Vol. 2, 1964, pp. 1380-1382.

H.J. Kelley, R.E. Kopp and H.G. Moyer, "Singular extremals", in Topics in Optimization, G. Leitmann, Ed., Academic Press, New York 1967.

H.J. Kelley, "A transformation approach to singular subarcs in optimal trajectory and control problems", *SIAM J. Control*, Vol. 2, 1964, pp. 234-240.

B.S. Goh, "Necessary conditions for singular extremals involving multiple control variables", *SIAM J. Control*, Vol. 4, 1966, pp. 716-731.

B.S. Goh, "The second variation for the singular Bolza problem", *SIAM J. Control*, Vol. 4, 1966, pp. 309-325.

J.B. Moore, "The singular solutions to a singular quadratic minimization problem", *Int. J. Control*, Vol. 20, 1974, pp. 383-393.

J.L. Speyer and D.H. Jacobson, "Necessary and sufficient conditions for optimality for singular control problems; a transformation approach", *J. Math. Anal. Appl.*, Vol. 33, 1971, pp. 163-187.

P.J. Moylan, "Preliminary simplifications in state-space impedance synthesis", *IEEE Trans. Circuits and Systems*, Vol. CAS-21, 1974, pp. 203-206.

R.E. O'Malley, Jr., and A. Jameson, "Singular perturbations and singular arcs - Part I", *IEEE Trans. Automatic Control*, Vol. AC-20, 1975, pp. 218-226.

Y.C. Ho, "Linear stochastic singular control problems", *J. Opt. Theory and Appl.*, Vol. 9, 1972, pp. 24-31.

D.J. Clements, B.D.O. Anderson and P.J. Moylan, "Matrix inequality solution to linear-quadratic singular control problems", *IEEE Trans. Automatic Control*, Vol. AC-22, 1977, pp. 55-57.

D.J. Clements and B.D.O. Anderson, "Transformational solution of singular linear-quadratic control problems", *IEEE Trans. Automatic Control*, Vol. AC-22, 1977, pp. 57-60.

V. Dolezal, "The existence of a continuous basis of a certain linear subspace of E_r which depends on a parameter", *Cas. Pro. Pest. Mat.*, Vol. 89, 1964, pp. 466-468.

[25] L. Weiss and P.L. Falb, "Dolezal's theorem, linear algebra with continuously parameterized elements and time-varying systems", *Math. Syst. Theory*, Vol. 3, 1969, pp. 67-75.

[26] F.R. Gantmaker, Theory of Matrices, Chelsea Pub. Co., New York, 1959.

[27] R. Courant and D. Hilbert, Methods of Mathematical Physics - Vol. II, Interscience Publishers Inc., New York, 1962.

[28] M.G. Wood, J.B. Moore and B.D.O. Anderson, "Study of an integral equation arising in detection theory", *IEEE Trans. Information Theory*, Vol. IT-17, 1971, pp. 677-686.

CHAPTER IV

DISCRETE-TIME LINEAR-QUADRATIC SINGULAR
CONTROL AND CONSTANT DIRECTIONS

INTRODUCTION

In continuous-time, the general linear-quadratic control problem is called singular
the weighting matrix of the controls in the integrand of the cost is singular any-
ere within the time interval of interest. In such cases, the Riccati equation nat-
lly associated with this problem is not well defined, so an alternative approach to
ving the problem is required.

In Chapter III, it is shown that a singular control problem can (subject to the
isfaction of certain structural and differentiability assumptions) be solved, using
solution of another control problem, again of a linear-quadratic nature and poss-
y singular, but of lower state and/or lower control space dimension. This reduction
cedure is continued until one of three possible terminating problems is obtained:
onsingular problem, or one with zero state dimension, or one with zero control
ension, each of which can be solved in a straightforward manner. The solution of
originally given control problem can be simply constructed from the solution of
terminating problem and knowledge of the reduction procedure.

Moreover, in carrying out any state dimension reduction, it becomes evident that
t of the optimal cost depends on the coefficient matrices of the problem, excluding
terminal weighting matrix, in an identifiable way.

In contrast, it is well known that for any discrete-time linear-quadratic optimal
trol problem, regardless of the singularity or otherwise of any matrices in the cost,
associated Riccati equation is well defined and may be solved in a straightforward
ner to yield a solution of the optimal control problem. As such, the concept of
gularity is apparently meaningless for discrete-time problems.

As mentioned above, one possible (though not necessary) consequence of singularity
continuous-time is the identification of blocks of the matrix defining the perform-
e index from knowledge of the coefficient matrices only. It is this idea that has
n studied in the context of the discrete-time linear-quadratic control problem and
associated Riccati equation in [1, 2]. There the concept of a constant direction
introduced, being a direction in which the solution of the Riccati equation is,
st, constant (except for a finite transient period), and second, independent of the
ninal weighting matrix. In [1], for the multi-input constant coefficient case for
estricted class of coefficient matrices, (the nature of the restriction being des-
bed in Section 3), the approach is to identify all constant directions of all
ers and then to reduce the dynamic order of the Riccati equation away from the
al transient period. Also introduced in [1] is the idea of a state being taken to
o optimally on an interval as a way of characterizing a constant direction. In

this paper, this idea also proves to be useful.

Work related to that in [1] was reported in [3] where the dual of the control problem, that of covariance factorization, is studied. These authors considered only scalar covariances, which corresponds to assuming a scalar input in the control problem formulation. Nonstationary covariances are studied, and to handle the time-varying case, the idea of a constant direction is extended and called a degenerate direction.

The contributions of this chapter are two-fold. Primarily, we study in detail the question of characterizing all constant directions for an arbitrary discrete-time linear-quadratic control problem. This generalizes corresponding results in [1] though our methods are somewhat different. Secondarily, we obtain results analogous to those for the general continuous-time singular linear-quadratic control problem. In particular, from the coefficient matrices we construct a matrix which plays the role of the control weighting matrix in continuous-time. That is, singularity of this matrix allows the originally given problem to be solved in terms of the solution of another control problem but with lower state and/or lower control dimension. Moreover, a reduction procedure which terminates in one of three control problems can be defined. Each of these terminating problems, a nonsingular problem, a zero state dimension problem and a zero control dimension problem can be solved simply; the solution to the originally given problem can be computed by tracing back through the reduction procedure.

An outline of the chapter is as follows. Section 2 reviews the general discrete-time control problem. In Section 3, we characterize j-constant directions completely in terms of being taken to zero optimally in j steps for some terminal weighting matrix and as vectors in the range of a certain matrix. Section 4 considers the problem of control space dimension reduction while in Section 5, state dimension reduction is demonstrated provided 1-constant directions exist. The results of the preceding sections are combined in Section 6 to derive an algorithm which solves any singular discrete-time control problem, making use of computational simplifications arising from constant direction existence. Section 7 contains brief concluding remarks, including mention of connections with the Silverman structure algorithm [4], and the recently developed "fast" methods for Riccati equation solution of Kailath and his coworkers, [5, 6].

The majority of the results in this chapter appeared in [7].

2.   LINEAR-QUADRATIC CONTROL IN DISCRETE TIME

In this section, we review basic results concerning the discrete-time, time-invariant, linear-quadratic control problem. On the interval [K,N] consider the dynamic system

$$x(i+1) = Ax(i) + Bu(i), \qquad i=K, \ldots, N-1 \qquad (2.1)$$
$$x(K) = x_K$$

ere x(i) is an n-dimension state vector, u(i) is an m-dimension control

ctor, and A and B are matrices with dimensions consistent with x and u. Let

U_K^{N-1} be a control sequence u(K), u(K=1), ..., u(N-1) and assign to the initial

ate x_K, control sequence U_K^{N-1} and terminal weighting matrix S, the cost funct-

nal

$$V_{N-K}[x_K, U_K^{N-1}, S] = x'(N)Sx(N) + \sum_{i+K}^{N-1} \{x'(i)Qx(i)$$

$$+ 2x'(i)Cu(i) + u'(i)Ru(i)\} \qquad (2.2)$$

ere x(K), x(K+1), ..., x(N) is the trajectory of (2.1) generated by U_K^{N-1} and

. The matrices Q, C, R and S are of appropriate dimension, with Q, R and S

mmetric. (No definiteness assumptions are yet made). Finally, define

$$v_{N-K}^*[x_K, S] = \inf_{U_K^{N-1}} V_{N-K}[x_K, U_K^{N-1}, S]. \qquad (2.3)$$

Consider a family of dynamics (2.1) and cost functionals (2.2) indexed by

0, 1, ..., N-1. The discrete-time linear-quadratic control problem can now be

ated in two parts.

1. Find necessary and sufficient conditions for
    $V_{N-K}[x_K, U_K^{N-1}, S]$ to be bounded below, independently
    of U_K^{N-1}, for each x_K and each K=0, ..., N-1.

2. If these conditions hold, determine $v_{N-K}^*[x_K, S]$ and, (2.4)
    if it exists, a control sequence U_K^{*N-1} depending on
    x_K such that

$$v_{N-K}^*[x_K, S] = V_{N-K}[U_K^{*N-1}, S]$$

    for each K=0, ..., N-1.

As a matter of terminology, we shall say that the control problem has a solution

[0, N] for terminal weighting matrix S whenever (2.4) has a solution.

The solution to this problem is well-known, and depends largely on the following

mma which we shall not prove here. The notations $N(X)$, $X^{\#}$, $X \geq 0(>0)$ denote

spectively null space, Moore-Penrose pseudo-inverse, nonnegative (positive definite-

ss of the matrix X.

Lemma IV.2.1: Consider the quadratic form $q(y, v) = y'Fy + 2y'Hv + v'Gv$ for

matrices F=F', G=G' and H and vectors y and v of arbitrary but consistent

dimensions, and define $q^*(y) = \inf_v q(y, v)$. The following three conditions are

equivalent:

(a) $q^*(y) > -\infty$ for each y

(b) $G \geq 0, N(G) \subseteq N(H)$

(c)   there exists a symmetric matrix   X   such that

$$\begin{bmatrix} F-X & H \\ H´ & G \end{bmatrix} \geq 0.$$

Moreover, if any one of the above conditions holds, then (c)   is satisfied by $X^* = F - HG^{\#}H´$. In addition, $X^* \geq X$ for any other X satisfying (c). Finally if for each y we set $v^* = -G^{\#}H´y$, then

$$q^*(y) = q(y, v^*) = y´X^*y.$$

Using this lemma and the next definitions, we shall be able to state the solution of (2.4).

<u>Definition</u>: The set of allowable weighting matrices, denoted by S, is the set of n×n symmetric matrices P such that $B´PB + R \geq 0$ and $N(B´PB + R) \subseteq N(A´PB + C)$. (The reason for this definition will become clear below).

<u>Remark 2.1</u>: The set S is in a sense open at one end. Specifically, if $S_0 \in S$ and $S \geq S_0$, then $S \in S$. To show this we must verify two properties of S, (a) $B´SB + R \geq 0$ and (b) $N(B´SB + R) \subseteq N(A´SB + C)$. Now (a) follows readily by writing $B´SB + R$ as $(B´S_0B + R) + B´(S - S_0)B$. For (b), assume $u \in N(B´SB + R)$. Then since $B´S_0B + R \geq 0$ and $S \geq S_0$, we conclude that $u \in N(B´S_0B + R)$, and $(S-S_0)Bu = 0$. However $S_0 \in S$. Therefore $u \in N(A´S_0B + C)$, and the result follows by writing $(A´SB + C)u = (A´S_0B + C)u + A´(S-S_0)Bu$.

We can now write down the solution to problem (2.4) as

<u>Theorem II.2.1</u>: The control problem has a solution on [0, N] for terminal weighting matrix   S   if and only if the   n×n   symmetric matrix P(i+1) $\in S$ for each i=0, ..., N-1, where P(i) is defined by the recursion

$$P(i) = A´P(i+1)A + Q$$

$$- [A´P(i+1)B + C][B´P(i+1)B + R]^{\#}[A´P(i+1)B + C]´,$$

$$i=0, ..., N-1 \tag{2.5}$$

$$P(N) = S.$$

If P(i) is so defined, then the control sequence U_K^{*N-1} defined by

$$u^*(i) = -[B´P(i+1)B + R]^{\#}[A´P(i+1)B + C]´x(i),$$

$$i=K, ..., N-1 \tag{2.6}$$

achieves the infimum for each K=0, ..., N-1. That is, with $U_K^{*N-1} = [u^*(K), ..., u^*(N-1)]$ we have

$$V^*_{N-K}[x_K, S] = V_{N-K}[x_K, U^{*N-1}_K, S] = x'_K P(K) x_K. \qquad (2.7)$$

oof: A simple application of Lemma 2.1 and the Principle of Optimality gives the
result.

marks 2.2: 1. Should $P(i+1) \notin S$ for some $i=0, \ldots, N-1$ then $V^*_{N-i}[x_i, S] = -\infty$
r at least one x_i.

    2. Note that $P(i)$ is also a function of the terminal weighting matrix
$\in S$. When we wish to make this dependence explicit, we shall write $P(i, S)$ for
i).

    3. If $B'P(i+1)B + R$ is nonsingular for each $i=0, \ldots, N-1$ then the
timal control sequence U^{*N-1}_0 defined by (2.6) is unique and so, therefore, is the
rresponding optimal trajectory of (2.1) for $K=0$ generated by U^{*N-1}_0 and x_0. If
r some $i=0, \ldots, N-1$, $B'P(i+1)B + R$ is singular, then the optimal control at time
is not uniquely defined, [i.e., alternatives to (2.6) are possible] and the optimal
ate $x^*(i+1)$ of the trajectory is also not unique. However, the definition of
(i) in (2.6) is the optimal control at time i of minimum norm, and as such is
iquely defined. For later reference, we call a control sequence U^{*N-1}_K with con-
ols defined by (2.6) the minimum norm optimal control sequence.

    4. If $P(i+1) \in S$, then the matrix $P(i)$ defined in equation (2.5)
y also be characterized as the maximal symmetric solution of

$$\begin{bmatrix} A'P(i+1)A + Q - P(i) & A'P(i+1)B + C \\ B'P(i+1)A + C' & B'P(i+1)B + R \end{bmatrix} \geq 0 . \qquad (2.8)$$

is fact will be used in the development of the state-space reduction procedure later
the chapter.

    The preceding discussion amounts to the standard approach for solving the linear-
adratic control problem, in that $P(N-i, S)$ is determined by separate sequential
himizations over each of the controls $u(N-1), u(N-2), \ldots, u(N-i)$ respectively.
wever, it is possible to calculate $P(N-i, S)$ directly by minimizing with respect
an extended control vector $u'_{(i)}(N-i) \triangleq [u'(N-i) \ldots u'(N-1)]$, provided we suit-
ly define the dynamics and the cost.

    In particular, consider $i = 2$ and the interval $[N-2, N]$. From (2.1) and (2.2)
$K = N-2$ we can write the dynamics as

$$x(N) = A_{(2)}x(N-2) + B_{(2)}u_{(2)}(N-2) \qquad (2.9)$$

re $A_{(2)} \triangleq A^2$ and $B_{(2)} \triangleq [AB \quad B]$ and the cost as

$$V_{(2)}[x(N-2), U^{N-1}_{N-2}, S] = x'(N)Sx(N) + \{x'(N-2)Q_{(2)}x(N-2)$$
$$+ 2x'(N-2)C_{(2)}u_{(2)}(N-2) + u'_{(2)}(N-2)R_{(2)}u_{(2)}(N-2)\}$$
$$\qquad (2.10)$$

where

$$Q_{(2)} \triangleq A'QA + Q$$

$$C_{(2)} \triangleq [A'QB + C \quad A'C]$$

$$R_{(2)} \triangleq \begin{bmatrix} B'QB + R & B'C \\ C'B & R \end{bmatrix}.$$

Consequently, if the control problem associated with (2.1) and (2.2) has a solution on [0, N] for terminal weighting S then P(N-2, S) is given by

$$P(N-2, S) = A'_{(2)}SA_{(2)} + Q_{(2)}$$

$$- [A'_{(2)}SB_{(2)} + C_{(2)}][B'_{(2)}SB_{(2)} + R_{(2)}]^{\#}[A'_{(2)}SB_{(2)} + C_{(2)}]'. \tag{2.11}$$

For convenience, we call the above the 2nd stage control problem, with terminal weighting S.

In general, suppose the jth stage control problem with terminal weighting S has been defined in terms of quantities $A_{(j)}$, $B_{(j)}$, $Q_{(j)}$, $C_{(j)}$, $R_{(j)}$ and $u_{(j)}(i)$; then the $(j+1)^{th}$ stage control problem with terminal weighting S is defined in terms of the quantities

$$A_{(j+1)} \triangleq A_{(j)}A, \quad B_{(j+1)} \triangleq [A_{(j)}B \quad B_{(j)}], \quad Q_{(j+1)} \triangleq A'Q_{(j)}A + Q$$

$$C_{(j+1)} \triangleq [A'Q_{(j)}B + C \quad A'C_{(j)}], \quad R_{(j+1)} \triangleq \begin{bmatrix} B'Q_{(j)}B + R & B'C_{(j)} \\ C'_{(j)}B & R_{(j)} \end{bmatrix} \tag{2.12}$$

$$u'_{(j+1)}(i) \triangleq [u'_{(j)}(i) \quad u'(i+j)].$$

Now it is clear that if the control problem associated with (2.1) and (2.2) has a solution on [0, N] then the jth stage control problem with terminal weighting P(i+j, S) defined by

$$x(i+j) = A_{(j)}x(i) + B_{(j)}u_{(j)}(i) \tag{2.13}$$

and

$$V_{(j)}[x(i), u_{(j)}(i), P(i+j, S)] = x'(i+j)P(i+j, S)x(i+j)$$

$$+ \{x'(i)Q_{(j)}x(i) + 2x'(i)C_{(j)}u_{(j)}(i)$$

$$+ u'_{(j)}(i)R_{(j)}u_{(j)}(i)\} \tag{2.14}$$

has a solution for all [i, i+j] ⊂ [0, N] and conversely. Moreover,

$$P(i, S) = A_{(j)}' P(i+j, S) A_{(j)} + Q_{(j)}$$

$$- [A_{(j)}' P(i+j, S) B_{(j)} + C_{(j)}][B_{(j)}' P(i+j, S) B_{(j)} + R_{(j)}]^{\#} [A_{(j)}' P(i+j, S) B_{(j)} + C_{(j)}]'$$

$$(2.15)$$

.th an optimal control $u_{(j)}^{*}(i)$ obviously equivalent to the optimal sequence
$i+j-1$. Moreover, existence of the optimal control implies the following replacement
the requirement that $P(i+1) \in S$:

$$N[B_{(j)}' P(i+j, S) B_{(j)} + R_{(j)}] \subset N[A_{(j)}' P(i+j, S) B_{(j)} + C_{(j)}]. \qquad (2.16)$$

istence also implies that $B_{(j)}' P(i+j, S) B_{(j)} + R_{(j)} \geq 0$ but we shall make more
e of (2.16).

By setting $i + j = N$, (2.15) becomes an equation for $P(N-j, S)$. Of course,
lculation of $P(N-j, S)$ using the jth stage formulation is highly inefficient.
vertheless, it is of theoretical interest, in that, as will later be evident, the
constant directions of the jth stage control problem are identical to the
constant directions of the original control problem. (See the next section for a
finition of j-constant directions).

CONSTANT DIRECTIONS - BASIC PROPERTIES

As stated in the introduction, we are interested in those directions in which
e Riccati equation (2.5) is completely determined by the coefficient matrices (save
r a possible final transient period) and at the same time is independent of the
rminal weighting matrix S. If such directions exist, the Riccati equation is
ally of lower dynamic order than it appears and solving (2.5) recursively involves
rrying out certain redundant calculations at each step of the recursion.

As in [1] we formalize the notion of constant directions in the following way:

Definition: Suppose $1 \leq j \leq N-1$. The n-vector α is called a j-constant dir-
tion of (2.5) on $[0, N]$ if and only if $P(i, S)\alpha$ is independent of those S in
for which a solution to the optimal control problem exists and of i for
$= 0, 1, \ldots, N-j$.

Denote the set of j-constant directions by I_j. Clearly, each I_j is a finite-
mensional linear space and is therefore completely described by a finite set of
nearly independent vectors. Moreover, for $1 \leq j \leq N-2$, it is immediate that
$+1 \supseteq I_j$.

It is of interest to have available various alternative characterizations of
constant directions, and our main task in this section is to develop these alternat-
e characterizations. One such characterization will involve the notion of a state
ing taken to zero optimally in j steps, which we now define precisely.

For each j with $1 \leq j \leq N-1$, consider the control problem with dynamics and cost as in (2.1) and (2.2), restricted to the interval [N-j, N], with coefficient matrices A, B, Q, C and R and terminal weighting matrix $\bar{S}$. Suppose that this control problem has initial state $x(N-j) = \alpha$.

Definition: We say that α can be taken to zero optimally in j steps for $S = \bar{S}$ if and only if there exists a control sequence U_{N-j}^{N-1} such that the corresponding trajectory of (2.1) satisfies $x(N-j) = \alpha$ and $x(N) = 0$, and $V_j[\alpha, U_{N-j}^{N-1}, \bar{S}]$ $= V_j^*[\alpha, \bar{S}]$. (An equivalent definition in terms of the jth stage control problems of (2.13) and (2.14) is obviously possible).

In [1], it is shown that the j-constant directions are completely characterized by the property of being taken to zero optimally in j steps for zero terminal weighting matrix when C=0, R=0, Q≥0, S≥0 and A is invertible. Moreover, the 1-constant directions can not only be characterized in this form but also simply as the range of the matrix $A^{-1}B$, while the j-constant directions are related to the null space of a matrix W_j (defined in [1]). The remainder of this section is concerned with the extensions of these results to control problems with coefficient matrices which are arbitrary, save that a solution to the control problem exists on [0, N] for some terminal weighting matrix $S \in \mathcal{S}$. Theorem 3.1 below contains the main result; it is preceded by several lemmas.

Define the (n+jm) × (n+jm) dimensional matrix

$$\Lambda_{(j)} = \begin{bmatrix} A_{(j)} & B_{(j)} \\ C'_{(j)} & R_{(j)} \end{bmatrix}. \tag{3.1}$$

This matrix proves a major tool in analyzing constant directions, and we show first that vectors in its nullspace define constant directions.

Lemma IV.3.1: Suppose $w_{(j)} \in N(\Lambda_{(j)})$ and partition $w'_{(j)}$ as $[\alpha'\ \beta'_{(j)}]$ where α is n-dimensional and $\beta_{(j)}$ is jm-dimensional. Then if the control problem has a solution on [0, N] for terminal weighting matrix S, we have

$$P(i, S)\alpha = Q_{(j)}\alpha + C_{(j)}\beta_{(j)} \quad \text{for each} \quad i = 0, 1, \ldots, N-j \tag{3.2}$$

so that α is a j-constant direction.

Proof: Since $w_{(j)} \in N(\Lambda_{(j)})$ we have

$$A_{(j)}\alpha + B_{(j)}\beta_{(j)} = 0 \quad \text{and} \quad C'_{(j)}\alpha + R_{(j)}\beta_{(j)} = 0. \tag{3.3}$$

Postmultiply (2.15) by α, and use (3.3) to obtain

$$i, S)\alpha = A'_{(j)}P(i+j, S)A_{(j)}\alpha + Q_{(j)}\alpha$$

$$-[A'_{(j)}P(i+j, S)B_{(j)} + C_{(j)}][B'_{(j)}P(i+j, S)B_{(j)} + R_{(j)}]^{\#}[A'_{(j)}P(i+j, S)B_{(j)} + C_{(j)}]'\alpha$$

$$= -A'_{(j)}P(i+j, S)B_{(j)}\beta_{(j)} + Q_{(j)}\alpha$$

$$+[A'_{(j)}P(i+j, S)B_{(j)} + C_{(j)}][B'_{(j)}P(i+j, S)B_{(j)} + R_{(j)}]^{\#}[B'_{(j)}P(i+j, S)B_{(j)} + R_{(j)}]\beta_{(j)}.$$

hypothesis, the control problem has a solution for terminal weighting S. There-
re conditions (2.16) hold for each [i, i+j] ⊂ [0, N] and so consequently does the
entity

$$A'_{(j)}P(i+j, S)B_{(j)} + C_{(j)} = [A'_{(j)}P(i+j, S)B_{(j)} + C_{(j)}][B'_{(j)}P(i+j, S)B_{(j)} + R_{(j)}]^{\#}$$

$$\times [B'_{(j)}P(i+j, S)B_{(j)} + R_{(j)}]. \quad (3.4)$$

Substituting (3.4) in the above we obtain (3.2).

marks 3.1: 1. Suppose that $x(N-j) = \alpha$ and $u_{(j)}(N-j) = \beta_{(j)}$. Then this control
uses $x(N) = 0$, in view of the first equation of (3.3). Further, the performance
dex $V_{(j)}[x(N-j), u_{N-j}^{N-1}, S]$ can be evaluated, using (2.14) with $i + j = N$ and the
ct that $x(N) = 0$, as

$$V_j[x(N-j), u_{N-j}^{N-1}, S] = \alpha'Q_{(j)}\alpha + 2\alpha'C_{(j)}\beta_{(j)} + \beta'_{(j)}R_{(j)}\beta_{(j)}$$

$$= \alpha'P(N-j, S)\alpha \cdot$$

e second equality comes from (3.2) and (3.3). Thus not only does α have the sig-
ficance of being a constant direction, but $\beta_{(j)}$ has the significance of being a
trol sequence which is optimal, and which drives $x(N-j)$ to $x(N) = 0$.

2. It is possible that there could exist two different vectors
$) = [\alpha' \quad \beta'_{(j)}]'$ and $\bar{w}_{(j)} = [\alpha' \quad \bar{\beta}'_{(j)}]'$ both in $N[\Lambda_{(j)}]$. Then $\beta_{(j)} - \bar{\beta}_{(j)}$
$N[B_{(j)}]'$; using this fact and (2.16), it follows that also $\beta_{(j)} - \bar{\beta}_{(j)} \in N[C_{(j)}]$.
such a circumstance, the same performance index results from using $\beta_{(j)}$ and $\bar{\beta}_{(j)}$.
it turns out, this nonuniqueness implies that there exist superfluous controls, a
int we explore more fully in the next section.

If the control problem has a solution on [0, N] for terminal weighting S_0 it
so has a solution on [0, N] for each terminal weighting S satisfying $S \geq S_0$.
s, from Lemma 3.1 and under the hypothesis of this lemma, we can assert that
$[\alpha, S] = V_k^*[\alpha, S_0]$ for each $S \geq S_0$ and any $k = j, \ldots, N$, with the optimum
ing achievable using the control sequence defined by $\beta_{(j)}$. The next lemma explores
consequences of such an equality of performance indices for a state α and some

Lemma IV.3.2: Suppose that $V_j^*[\alpha, S] = V_j^*[\alpha, S_0]$ for a state α and all
$S \geq S_0$, for some $S_0 \in S$. Then for $\bar{S} > S_0$, with the overbar indicating the

terminal weighting matrix $\bar{S}$ is being considered, we have

(a) $\bar{x}^*(N) = 0$ for <u>any</u> optimal control sequence $\bar{U}_{N-j}^{*N-1}$ associated with $\bar{S}$,

(b) $v_j^*[\alpha, S] = v_j[\alpha, \bar{U}_{N-j}^{*N-1}, S] = v_j^*[\alpha, S_0]$ for all $S \geq S_0$.

<u>*Proof*</u>: By hypothesis, $v_j^*[\alpha, \bar{S}] = v_j^*[\alpha, S] \leq v_j[\alpha, U_{N-j}^{N-1}, S_0]$ for any control sequence U_{N-j}^{N-1}, and therefore for any optimizing control sequence $\bar{U}_{N-j}^{*N-1}$ associated with terminal weighting $\bar{S}$. Now,

$$v_j^*[\alpha, \bar{S}] = v_j[\alpha, \bar{U}_{N-j}^{*N-1}, S_0] + \bar{x}^{*\prime}(N)(\bar{S}-S_0)\bar{x}^*(N).$$

Consequently, $\bar{x}^*(N) = 0$ since $\bar{S} > S_0$. This proves (a). Part (b) follows trivially.

Part (a) and (b) of the above lemma yield:

<u>Corollary 3.1</u>: Suppose that $v_j^*[\alpha, S] = \overset{*}{\underset{j}{v}}[\alpha, S_0]$ for a state α and all $S \geq S_0$ for some $S_0 \in S$. Then α can be taken to zero optimally in j steps for all $S \geq S_0$.

It is important to note that in corollary 3.1 the minimum norm optimal control is not guaranteed to take α to zero optimally in j steps for $S \geq S_0$ but $S \not= S_0$. This does not invalidate the result of Corollary 3.1, however, because Definition 3.2 only requires <u>some</u> optimal control to take α to zero in j steps, not necessarily the minimum norm control.

By way of a converse to this corollary, we have the following:

<u>Lemma IV.3.3</u>: Suppose that α can be taken to zero optimally in j steps for some $S_0 \in S$. Then α can be taken to zero optimally in j steps for all $S \geq S_0$ and $v_j^*[\alpha, S] = v_j^*[\alpha, S_0]$.

<u>*Proof*</u>: It is easily checked that the control taking α to zero optimally for $S_0 \in S$ also is optimum for all $S \geq S_0$.

We have shown that the null vectors of $\Lambda_{(j)}$ determine states which can be taken to zero optimally. Let us now check the reverse of this idea.

<u>Lemma IV.3.4</u>: Suppose that α can be taken to zero optimally in j steps for all $S \geq S_0$ where S_0 is some element of S. Then there exists a vector $\beta_{(j)}$ such that $w_{(j)} \in N(\Lambda_{(j)})$ for $w_{(j)}' = [\alpha' \quad \beta_{(j)}']$.

<u>*Proof*</u>: With terminal weighting matrix $\bar{S} > S$, Lemma 3.2 implies that α can be taken to zero optimally in j steps with <u>any</u> optimizing control U_{N-j}^{*N-1} or equivalently $u_{(j)}^*(N-j)$. In particular, one such optimizing control is the minimum norm control

$$\bar{u}_{(j)}^*(N-j) = -[B_{(j)}'\bar{S}B_{(j)} + R_{(j)}]^{\#}[A_{(j)}'\bar{S}B_{(j)} + C_{(j)}]'\alpha \qquad (3.5)$$

and therefore since $\bar{x}^*(N) = 0$ we have

$$A_{(j)}\alpha - B_{(j)}[B_{(j)}'\bar{S}B_{(j)} + R_{(j)}]^{\#}[A_{(j)}'\bar{S}B_{(j)} + C_{(j)}]'\alpha = 0. \qquad (3.6)$$

w, $\bar{S} \in S$ and so the identity (3.4) holds for $P(i+j, S) = \bar{S}$, which together with
.6) implies

$$C_{(j)}^{\prime}\alpha - R_{(j)}[B_{(j)}^{\prime}\bar{S}B_{(j)} + R_{(j)}]^{\#}[A_{(j)}^{\prime}\bar{S}B_{(j)} + C_{(j)}]^{\prime}\alpha = 0. \qquad (3.7)$$

nce, with $\beta_{(j)} = \bar{u}_{(j)}^*(N-j)$, (3.6) and (3.7) become $A_{(j)}\alpha + B_{(j)}\beta_{(j)} = 0$ and
$_{j})^{\alpha} + R_{(j)}\beta_{(j)} = 0$, completing the proof of the lemma.

We can summarize the preceding results as follows:

Theorem IV.3.1: Suppose that the solution to the control problem exists on
[0, N] for some terminal weighting S. Then the following statements are equivalent.

(a) α is a j-constant direction of (2.5) on [0, N].

(b) α can be taken to zero optimally in j steps for all $S \in S$.

(c) There exists $w_{(j)} \in N(\Lambda_{(j)})$ with $w_{(j)}^{\prime} = [\alpha^{\prime} \quad \beta_{(j)}^{\prime}]$.

(d) [restricted form of (a)]. For some $S_0 \in S$ and all $S \geq S_0$,
$v_j^*[\alpha, S] = v_j^*[\alpha, S_0]$.

(e) [restricted form of (b)]. α can be taken to zero optimally in j steps
for some $S_0 \in S$.

reover, should any of the above hold, any optimal control associated with an $S \in S$
ch that $S - \eta I \in S$ for some $\eta > 0$ takes α to zero.

oof: The implications (a) $\Rightarrow$ (b), (e) $\Rightarrow$ (d) $\Rightarrow$ (c) $\Rightarrow$ (a) follow from
rollary 3.1, Lemma 3.3, Lemma 3.4 with Corollary 3.1, and Lemma 3.1 respectively.
nally, (b) $\Rightarrow$ (e) is trivial. The final part of the theorem is a consequence of
mma 3.2. This completes the proof.

We also have the following simple consequence of parts (a) and (c) of Theorem
1 and Remark 3.1.2.

Theorem IV.3.2: Suppose the control problem has a solution on [0, N] for some
terminal weighting S. Then the space of j-constant directions I_j is the
range of W_{j1}, where $W_j = [W_{j1}^{\prime} \quad W_{j2}^{\prime}]^{\prime}$ is a basis matrix for $N(\Lambda_{(j)})$. Moreover,
the dimension of I_j equals $s_j - p_j$ where s_j is the nullity of $\Lambda_{(j)}$ and
p_j is the nullity of $[B_{(j)}^{\prime} \quad R_{(j)}^{\prime}]^{\prime}$.

In this section, we have not singled out 1-constant directions for special atten-
on. This we shall do in later sections, since it is the 1-constant directions that
e most easily found [clearly $N(\Lambda_{(1)})$ is easier to compute than $N(\Lambda_{(j)})$] and, as
turns out, j-constant directions can be found by computing 1-constant directions
r a collection of problems.

Our next immediate task however is to analyse the issue of superfluous controls.

4. CONTROL SPACE DIMENSION REDUCTION

In this and the next section, we shall concentrate on the situation where 1-constant directions exist. In the previous section we showed in Theorem 3.2 that the number of linearly independent 1-constant directions ℓ is equal to the dimension s of $N(\Lambda)$ less the dimension p of the null-space of $[B' \quad R']'$. In this section, we argue that if $\ell < s$ or $p > 0$, then we can find a basis of the control space such that the dynamics (2.1) and the cost (2.2) are independent of p components of the control vector u. Consequently the solution of the control problem on $[0, N]$ can be shown to be equivalent to the solution of a control problem on $[0, N]$ with the same state basis but of lower control space dimension.

For $p = 0$, no reduction of the control space dimension will be possible. For the present, assume that $0 < p < m$. We show that the dimension of the control space can be reduced by p. (The case $p \geq m$ will be treated subsequently). Let W be a basis matrix for $N(\Lambda)$ and partition $\Lambda W = 0$ as

$$
\begin{array}{l} n \{ \\ m-p \{ \\ p \{ \end{array}
\begin{bmatrix} A & B_1 & B_2 \\ C_1' & R_{11} & R_{12} \\ C_2' & R_{12}' & R_{22} \end{bmatrix}
\begin{bmatrix} W_{11} & W_{12} \\ W_{21} & W_{22} \\ W_{31} & W_{32} \end{bmatrix}
=
\begin{bmatrix} 0 & 0 \\ 0 & 0 \\ 0 & 0 \end{bmatrix}
\qquad (4.1)
$$
$$
\underbrace{}_{n} \ \underbrace{}_{m-p} \ \underbrace{}_{p} \ \underbrace{}_{\ell} \ \underbrace{}_{p}
$$

with the dimensions of the various submatrices as shown.

With a basis change of $N(\Lambda)$ via column operations and a basis change of the control space via row operations on the final m rows of W, we transform W to a more suitable form. Specifically, since $[W_{11} \quad W_{12}]$ is an $n \times s$ matrix of column rank ℓ there exists an $s \times s$ nonsingular matrix T such that $[W_{11} \quad W_{12}]T = [\bar{W}_{11} \quad 0]$ with $\bar{W}_{11}$ of full column rank ℓ. With T change the basis of $N(\Lambda)$ to $\bar{W} = WT$ where

$$
\bar{W} = \begin{bmatrix} \bar{W}_{11} & 0 \\ \bar{W}_{21} & \bar{W}_{22} \\ \bar{W}_{31} & \bar{W}_{32} \end{bmatrix}.
$$

Since $\bar{W}$ has full column rank, so must $[\bar{W}_{22}' \quad \bar{W}_{32}']'$. Therefore there exists an $m \times m$ nonsingular matrix U such that $U[\bar{W}_{22}' \quad \bar{W}_{21}']' = [0 \quad I_p]'$. With U, change the control space basis. Dropping the bar notation, we have a basis of the control space such that a basis matrix W of $N(\Lambda)$ is given by

$$
W = \begin{bmatrix} W_{11} & 0 \\ W_{21} & 0 \\ W_{31} & I_p \end{bmatrix}.
\qquad (4.2)
$$

s result might also be obtained by directly constructing a basis of $N(\Lambda)$ as
$\left[\begin{matrix}\alpha_\ell\\\beta_\ell\end{matrix}\right], \ldots, \left[\begin{matrix}0\\\beta_{\ell-1}\end{matrix}\right], \ldots, \left[\begin{matrix}0\\\beta_s\end{matrix}\right]$. That such a basis exists follows from the fact
t $s = \ell + p$.

From the special form of W, we see that $\Lambda W = 0$ implies

$$B_2 = 0 \quad, \quad R_{12} = 0 \quad, \quad R_{22} = 0. \tag{4.3}$$

Assume that the control problem has a solution on $[0, N]$ for terminal weighting
rix S. Then because $N[B'SB + R] \subset N[A'SB + C]$ and every m–vector of the form
$v']'$ with v a p–vector lies in $N[B'SB + R]$, it follows that $C'v = 0$.
ause v is arbitrary, $C_2 = 0$.

Now partition the controls u as $[u_1'\ u_2']'$ where u_2 has dimension p. The
amics (2.1) then become

$$x(i+1) = Ax(i) + B_1u_1(i) \quad i = 0, \ldots, N-1 \tag{4.4}$$

$$x(0) = x_0$$

the cost (2.2) becomes

$$V[x_0, U_0^{N-1}, S] = x'(N)Sx(N)$$
$$+ \sum_{i=0}^{N-1} \{x'(i)Qx(i) + 2x'(i)C_1u_1(i) + u_1'(i)R_{11}u_1(i)\}. \tag{4.5}$$

ce, the control problem with dynamics (2.1) and cost (2.2) has a solution on
N] if and only if the control problem with dynamics (4.4), cost (4.5) and control
ce dimension m–p has a solution on $[0, N]$. Moreover, the solution of these two
trol problems are simply related, see Lemma 4.1 below.

It remains to consider the possibility $p = s - \ell \geq m$. Since p is the dimension
the nullspace of $[B'\ R]'$ and $[B'\ R]'$ has m columns, we must have $p = m$
$B = 0$, $R = 0$. Minor modification of the argument applying for $p < m$ yields
0 (assuming an optimal solution exists). We see then that the control vector u
no effect on the dynamics (2.1) or the cost (2.2); the control problem is trivial
$[0, N]$.

The results of the section are summarized as

Lemma IV.4.1: Assume that the dimension of $N(\Lambda)$ is s, that $N(\Lambda)$ has basis
matrix W with first n rows W_1, that the rank of W_1 is ℓ; define
$p = s - \ell \geq 0$. Then the control problem has a solution on $[0, N]$ for terminal
weighting matrix S if and only if a control problem of identical form but with
control dimension m–p has a solution on $[0, N]$ for terminal weighting matrix
S. Moreover, the solutions $P(i, S)$, $i = 0, \ldots, N-1$ of the associated Riccati
equations are identical while there exists a basis of the original control space
such that any optimizing control at time i, $u^*(i)$ is given by $[u_1^{*'}(i)\ u_2^{*'}(i)]'$

where $u_1^*(i)$ is any optimizing control of the lower control dimension problem and $u_2^*(i)$ is chosen arbitrarily. In particular, for $u^*(i)$ the minimum norm optimal control, $u_1^*(i)$ is the corresponding minimum norm optimal control for the problem of lower control dimension and $u_2^*(i) = 0$.

It is clear from this lemma that when $p > 0$, the amount of computation involved in solving the Riccati equation for the original control problem can be reduced to that involved in solving one of the same dynamic order but of lower control dimension. This lemma is analogous to the result that holds for continuous-time singular linear-quadratic control problems, as discussed in the previous chapter.

5.   STATE SPACE DIMENSION REDUCTION

In this section, we assume that the matrix $[B^{\prime} \ R]^{\prime}$ has full column rank or, equivalently, all superfluous controls have been eliminated. Then we know that the dimension of I_1 equals the nullity of $\Lambda = \Lambda_{(1)}$; let this number be ℓ. We will now show that if there are nontrivial 1-constant directions, i.e. $\ell > 0$, then the state space dimension can be reduced by ℓ, and a related control problem can be defined on the interval $[0, N-1]$ rather that $[0, N]$. Moreover, the solution of the Riccati equation and the optimal controls on $[0, N]$ are simply related to those for the reduced state dimension problem on $[0, N-1]$.

The first stage in the procedure is to choose a basis of the state space to display the constant parts of the matrices $P(i, S)$, $i = 0, \ldots, N-1$. Choose as a basis of the state space $\{\alpha_1, \ldots, \alpha_n\}$ arbitrarily, save that $\{\alpha_{n-\ell+1}, \ldots, \alpha_n\}$ spans I_1. With this state space basis, we have

$$P(i, S) = \begin{array}{c} \\ \\ \ell\{ \end{array} \begin{bmatrix} P_{11}(i, S) & P_{12} \\ P_{12}^{\prime} & P_{22} \end{bmatrix} \qquad i = 1, \ldots, N-1 \qquad (5.1)$$

where P_{12} and P_{22} are constant matrices independent of S and $i = 1, \ldots, N-1$.

By our assumption that $[B^{\prime} \ R]^{\prime}$ has full column rank, we know that for each α_i, $i = n-\ell+1, \ldots, n$ there exists a unique β_i such that $w_i = [\alpha_i^{\prime} \ \beta_i^{\prime}]^{\prime} \in N(\Lambda)$. Let Z be the matrix $[\beta_{n-\ell+1}, \ldots, \beta_n]$. Partition Q and C conformably with $P(i, S)$, i.e. set

$$Q = \begin{bmatrix} Q_{11} & Q_{12} \\ Q_{12}^{\prime} & Q_{22} \end{bmatrix} \quad \text{and} \quad C = \begin{bmatrix} C_1 \\ C_2 \end{bmatrix}. \qquad (5.2)$$

In this state space basis the result of Lemma 3.1 can now be restated as

$$P_{12} = Q_{12} + C_1 Z \quad \text{and} \quad P_{22} = Q_{22} + C_2 Z . \qquad (5.3)$$

Thus, we have completely identified P_{12} and P_{22} as constant parts of $P(i, S)$, $i = 0, \ldots, N-1$. Moreover, the theory developed in Section 3 says that no part of

$_1(i, S)$ is independent of $i = 0, \ldots, N-1$ and $S \in \mathcal{S}$, though some part may be
$_r$ $i = 0, \ldots, N-2$. (If some part of $P_{11}(i, S)$ were independent for $i = 0, \ldots, N-1$
$\in \mathcal{S}$, there would exist a vector in $N(\Lambda)$ not in I_1 and this would be a contra-
ction).

By virtue of (5.3), the evaluation of $P(i, S)$ via the Riccati equation (2.5)
$[0, N]$ clearly involves a substantial amount of unnecessary calculation. It
ld be of interest if we could show that $P_{11}(i, S)$, $i = 0, \ldots, N-1$ could be
mputed via a Riccati equation for P_{11} rather that P, presumably involving
fferent A, B, Q, C and R. Recall that the solution to the Riccati equation (3.5)
the maximal symmetric matrix $P(i, S)$ such that

$$
\begin{bmatrix}
A'P(i+1, S)A + Q - P(i, S) & A'P(i+1, S)B + C \\
B'P(i+1, S)A + C' & B'P(i+1, S)B + R
\end{bmatrix} \geq 0
\tag{5.4}
$$

$_r$ each $i = 0, \ldots, N-1$ with $P(N, S) = S$. With A partitioned as $A = [A_1 \ A_2]$
is can then be written as

$$
\begin{bmatrix}
A_1'P(i+1, S)A_1 & A_1'P(i+1, S)A_2 & A_1'P(i+1, S)B \\
\quad + Q_{11} - P_{11}(i, S) & \quad + Q_{12} - P_{12} & \quad + C_1 \\
A_2'P(i+1, S)A_1 & A_2'P(i+1, S)A_2 & A_2'P(i+1, S)B \\
\quad + Q_{12}' - P_{12}' & \quad + Q_{22} - P_{22} & \quad + C_2 \\
B'P(i+1, S)A_1 & B'P(i+1, S)A_2 & B'P(i+1, S)B \\
\quad + C_1' & \quad + C_2' & \quad + R
\end{bmatrix} \geq 0.
\tag{5.5}
$$

nce the vectors w_i form a basis matrix of $N(\Lambda)$ we also have the relations

$$
A_2 + BZ = 0 \quad \text{and} \quad C_2' + RZ = 0.
\tag{5.6}
$$

$e$multiply (5.5) by the nonsingular matrix

$$
T = \begin{bmatrix}
I & 0 & 0 \\
0 & I & Z \\
0 & 0 & I
\end{bmatrix}
$$

d postmultiply it by T'. Using (5.3) and (5.6), we obtain in this way another
$trix$ inequality which is equivalent to (5.5):

$$
\begin{bmatrix}
A_1'P(i+1, S)A_1 + Q_{11} & 0 & A_1'P(i+1, S)B + C_1 \\
\quad - P_{11}(i, S) & & \\
0 & 0 & 0 \\
B'P(i+1, S)A_1 + C_1' & 0 & B'P(i+1, S)B + R
\end{bmatrix} \geq 0
$$

which is of course in turn equivalent to

$$\begin{bmatrix} A_1'P(i+1, S)A_1 + Q_{11} - P_{11}(i, S) & A_1'P(i+1, S)B + C_1 \\ B'P(i+1, S)A_1 + C_1' & B'P(i+1, S)B + R \end{bmatrix} \geq 0. \qquad (5.7)$$

It is now evident that $P_{11}(i, S)$ is the maximal solution of the inequality (5.7). However we know that an equivalent definition is provided by a matrix Riccati equation

$$\hat{P}(i, S) = \hat{A}'\hat{P}(i+1, S)\hat{A} + \hat{Q}$$

$$- [\hat{A}'\hat{P}(i+1, S)\hat{B} + \hat{C}][\hat{B}'\hat{P}(i+1, S)\hat{B} + \hat{R}]^{\#}[\hat{A}'\hat{P}(i+1, S)\hat{B} + \hat{C}]'$$

$$i = 0, \ldots, N-2 \qquad (5.8)$$

where $\hat{P}(i, S) = P_{11}(i, S)$, $\hat{A} = \hat{A}_{11}$ (supposing A_1 is partitioned as $[\hat{A}_{11} \ \hat{A}_{21}]'$) and the other hat quantities are defined in terms of P_{21}, P_{22} and the original coefficient matrices A, B, Q, C and R. (The precise definitions are contained in the Appendix). Finally, we initialize (5.8) with

$$\hat{S} = \hat{P}(N-1, S) = P_{11}(N-1, S).$$

Let us summarize what we have shown so far. If the Riccati equation (2.5) has a solution on [0, N] for terminal weighting matrix S, and if the null space of Λ has dimension equal to ℓ, (or equivalently, surplus controls have been eliminated) then, modulo state and control space basis changes, the solution of (2.5) on [0, N] with P(N) = S is equivalent to the solution of (5.8) on [0, N-1] with $\hat{P}(N-1, S)$ = $P_{11}(N-1, S)$ and with P_{12} and P_{22} defined by (5.3) for $i = 0, \ldots, N-1$.

To complete this section, we point out that (5.8) can be associated with a control problem, closely related to and of the same form as that originally given, but now involving hat quantities and defined on [0, N-1] rather than on [0, N]. (This observation allows the relation of optimal controls for the two problems). Suppose that the state and control bases are chosen as described at the start of this section, and partition the state variable x as $[x_1' \ x_2']'$ with x_2 of dimension ℓ. Define new state and control variables

$$\hat{x} = x_1 \qquad (5.9)$$

$$\hat{u} = u - Zx_2.$$

With this notation, it follows from the dynamics of the original system (2.1), from the definitions of $\hat{A}$, $\hat{B}$ and from (5.3) and (5.6) that

$$\hat{x}(i+1) = \hat{A}\hat{x}(i) + \hat{B}\hat{u}(i), \quad i = 0, \ldots, N-2 \qquad (5.10)$$

$$\hat{x}(0) \quad = x_1(0) = \hat{x}_0.$$

uation (5.10) constitutes the dynamics of a reduced system on $[0, N-1]$.

In terms of the hat quantities, it is also possible after some manipulation to
ite

$$V[x_0, U_0^{N-1}, S] = \hat{V}[\hat{x}_0, \hat{U}_0^{N-2}, \hat{S}] + x'(0)\bar{P}x(0)$$

$$+ [u(N-1) + (B'SB+R)^{\#}(A'SB+C)'x(N-1)]'(B'SB+R)$$

$$\times [u(N-1) + (B'SB+R)^{\#}(A'SB+C)'x(N-1)] \qquad (5.11)$$

ere

$$\hat{V}[\hat{x}_0, \hat{U}_0^{N-2}, \hat{S}] \triangleq \hat{x}'(N-1)\hat{S}\hat{x}(n-1)$$

$$+ \sum_{i=0}^{N-2} \{\hat{x}'(i)\hat{Q}\hat{x}(i) + 2\hat{x}'(i)\hat{C}\hat{u}(i) + \hat{u}'(i)\hat{R}\hat{u}(i)\} \qquad (5.12)$$

d

$$\bar{P} = \begin{bmatrix} 0 & Q_{12} + C_1 Z \\ Q'_{12} + Z'C'_1 & Q_{22} + C_2 Z \end{bmatrix}. \qquad (5.13)$$

Now the right side of (5.11) is the sum of three terms, the first $\hat{V}[\hat{x}_0, \hat{U}_0^{N-2}, \hat{S}]$
ends on $\hat{u}(i)$, $i = 0, \ldots, N-2$ and $\hat{x}_0$, the second is constant and depends on
alone, while the third is a function of $u(N-1)$ and $x(N-1)$. The independence
these terms together with the nonnegativity of $B'SB + R$ allows us to conclude
at $V^*[x_0, S]$ is finite if and only if $\hat{V}^*[\hat{x}_0, \hat{S}]$ is finite, with these quantities
isfying

$$V^*[x_0, S] = \hat{V}^*[\hat{x}_0, \hat{S}] + x'(0)\bar{P}x(0). \qquad (5.14)$$

Altogether then we can solve the control problem on the interval $[0, N]$ in terms
another control problem of lower state space dimension on the interval $[0, N-1]$,
never there exist nontrivial 1-constant directions. [Knowing $\hat{u}(0)$, $x_1(0)$, $x_2(0)$,
obtain $u(0)$ from (5.9) and $x_1(1)$, $x_2(1)$ from $Ax(0) + Bu(0)$; then knowing
), $x_1(1)$, $x_2(1)$, we obtain $u(1)$ from (5.9), etc.].

We summarize the main result as

Theorem IV.5.1: Assume that the dimension of $N(\Lambda)$ is ℓ and that for any
basis matrix W, the rank of the matrix W_1 from the first n rows of W is
ℓ. Then the original control problem has a solution on $[0, N]$ for terminal
weighting matrix $S \in S$ if and only if the control problem with dynamics (5.10)
and cost (5.12) has solution on $[0, N-1]$ for terminal weighting matrix $\hat{S}$,
with the state dimension $n-\ell$. Moreover, the optimal costs are related as in
(5.14) and any optimizing control sequence U_0^{*N-1} can be related to an optimizing
sequence U_0^{*N-2}, using x_0 and noting (5.9).

6. TOTAL REDUCTION OF THE PROBLEM

In the last two sections, we have shown how the complexity of a linear-quadratic problem can be reduced in case there are 1-constant directions. In this section, we shall study what happens when there are j-constant directions for $j > 1$. Our principle conclusion is that repeated application of the reduction procedures applicable for 1-constant directions will ultimately eliminate all constant directions of any index. In establishing this conclusion, we shall draw heavily on the general theory of Section 3, as well as the procedures of the last two sections.

To begin with, we shall assume that the set of constant directions of a problem includes 1-constant directions. (Later, we shall show that if there are j-constant directions for $j > 1$, there must also be 1-constant directions. So it transpires that there is really no loss of generality in this assumption). Also, we shall assume that redundant controls have been eliminated. These assumptions mean that we can carry out the reduction procedure of the last section. The first lemma considers the effect of this procedure on the j-constant directions for $j > 1$.

Lemma IV.6.1: Suppose that the solution to the control problem exists on $[0, N]$ for some terminal weighting S. Assume that the state coordinate base is chosen such that the reduction procedure of Section 5 may be applied. Then with $j \geq 2$, α is a j-constant direction of the Riccati equation (2.5) if and only if α_1 is a (j-1)-constant direction of the Riccati equation (5.8), where $\alpha = [\alpha_1' \ \alpha_2']'$, α_2 having dimension ℓ.

Proof: Suppose that α_1 is a (j-1)-constant direction of the Riccati equation (5.8) for $j \geq 2$. Then, from the definition of a constant direction, and noting that (5.8) is defined on $[0, N-1]$, we have

$$\hat{P}(N-i, \hat{S})\alpha_1 = \text{constant} \tag{6.1}$$

independent of the weighting matrices $S \in \mathcal{S}$, for which a solution to a reduced problem exists and all $i \geq j$. Then for $\alpha = [\alpha_1' \ \alpha_2']'$ for any α_2 and any S for which the original problem has a solution

$$P(N-i, S)\alpha = \begin{bmatrix} P_{11}(N-i, S) & P_{12} \\ P_{12}' & P_{22} \end{bmatrix} \begin{bmatrix} \alpha_1 \\ \alpha_2 \end{bmatrix}$$

$$= \begin{bmatrix} \hat{P}(N-i, \hat{S}) & P_{12} \\ P_{12}' & P_{22} \end{bmatrix} \begin{bmatrix} \alpha_1 \\ \alpha_2 \end{bmatrix} \tag{6.2}$$

where $\hat{S}$ is of the form $P_{11}(N-1, S)$. Therefore from (6.1), which applies to all $\hat{S} \in \hat{\mathcal{S}}$ and a fortiori to those of the form $P_{11}(N-1, S)$,

$$P(N-i, S)\alpha = \text{constant} \tag{6.3}$$

all $S \in S$ and $i \geq j$. Thus, α is a j-constant direction of (2.5).

Conversely, suppose that α is a j-constant direction of (2.5) for $j \geq 2$.
te $\alpha' = [\alpha_1' \; \alpha_2']$. Then (6.3) holds for any $i \geq j$ and $S \in S$. Therefore, (6.1)
ds for any $i \geq j$ and $\hat{S} \in \hat{S}$ of the form $P_{11}(N-1, S)$. By Theorem 3.1, it is
'ficient to show that (6.1) holds for all $i \geq j$ and all $\hat{S} \geq \hat{S}_0$ for some $\hat{S}_0 \in \hat{S}$.

First, we show that if $S > S_0 \in S$, then $P_{11}(N-1, S) > P_{11}(N-1, S_0)$. We argue
contradiction. Suppose that there exists an $x \neq 0$ such that

$$P_{11}(N-1, S)x = P_{11}(N-1, S_0)x. \qquad (6.4)$$

any $\bar{S}$ satisfying $S_0 \leq \bar{S} \leq S$, we have $P_{11}(N-1, S_0) \leq P_{11}(N-1, \bar{S}) \leq P_{11}(N-1, S)$
, therefore $0 \leq x'[P_{11}(N-1, \bar{S}) - P_{11}(N-1, S_0)]x \leq x'[P_{11}(N-1, S) - P_{11}(N-1, S_0)]x$
'. Hence, $P_{11}(N-1, \bar{S})x = P_{11}(N-1, S_0)x$ for all $\bar{S}$ such that $S_0 \leq \bar{S} \leq S$. Now
.ce $S > S_0$, an argument similar to that in Lemmas 3.2 and 3.3 shows that
$(N-1, \bar{S})x = P_{11}(N-1, S_0)x$ for any $\bar{S} \geq S_0$, even if $\bar{S} \leq S$ does not hold. There-
e (6.2) for i=1 and Theorem 3.1 imply that $[x' \; \alpha_2']'$ is a 1-constant direction,
ch is a contradiction. (In view of the basis chosen, all 1-constant directions
e the form $\alpha = [0 \; \alpha_2']'$).

Thus, (6.1) holds for $\hat{S}_0 = P_{11}(N-1, S_0)$ and $\hat{S} = P_{11}(N-1, S)$ with $\hat{S} > \hat{S}_0$.
in, an argument as in Lemmas 3.2 and 3.3 implies that (6.1) holds for all $\hat{S} \geq \hat{S}_0$.
refore, by Theorem 3.1, α_1 is a (j-1)-constant direction of (4.7) for $j \geq 2$.
s completes the proof of the lemma.

Hence, if the space of 1-constant directions is non-zero, the state space
uction procedure holds and the j-constant directions, $j \geq 2$, of the originally
en problem become (j-1)-constant directions of the reduced state dimension problem.
sider now the repeated application of the idea of the above lemma. Suppose that a
uced problem is obtained via the procedures of the last two sections.

Now if this new problem has $N(\Lambda)$ nonempty, we first eliminate any unnecessary
trols by the procedure of Section 4. Then if $\hat{I}_1 \neq \{0\}$, with $\hat{I}_1$ the space of
onstant directions for the reduced state dimension problem, we can again reduce
ording to Section 5. Clearly, if at some stage in the above procedure we obtain
$= \{0\}$ for one of the reduced problems, we cannot proceed any further. Now by
ma 6.1 this is equivalent to having $I_\ell = I_{\ell+1}$ for some ℓ in the original
olem. We will now show that $I_\ell = I_{\ell+1}$ for some ℓ implies that $I_\ell = I_{\ell+k}$
every $k \geq 2$ in the original problem. This means that there is no way, possibly
ng some other algorithm than that presented, of eliminating further constant
ections.

Since $I_{\ell+1} \subset I_{\rho+k}$ for all $k \geq 2$ we could only have $I_{\ell+1} \neq I_{\ell+k}$ if there
$\ni$ (k+ℓ)-constant directions which were not (ℓ+1)-constant directions for the
ginal problem, or k-constant directions which were not 1-constant directions for
reduced problem. Since the reduced problem has no 1-constant direction, the
llt will follow from the following lemma:

Lemma 6.2: If $I_1 = \{0\}$, then $I_k = \{0\}$ for all $k > 1$.

Proof: We argue by contradiction. Let j be the least value of $k > 1$ for which $I_j \neq \{0\}$. Let $w_{(j)} = [\alpha' \ \beta'_{(j)}]' \in N[\Lambda_{(j)}]$ with $\alpha \neq 0$. By (2.12),

$$\begin{bmatrix} A_{(j-1)}A & A_{(j-1)}B & B_{(j-1)} \\ B'Q_{(j-1)}A + C' & B'Q_{(j-1)}B + R & B'C_{(j-1)} \\ C'_{(j-1)}A & C'_{(j-1)}B & R_{(j-1)} \end{bmatrix} \begin{bmatrix} \alpha \\ \beta \\ \gamma \end{bmatrix} = 0 \qquad (6.5)$$

where $\beta_{(j)} = [\beta' \ \gamma']'$. Immediately, we see that

$$\Lambda_{(j-1)} \begin{bmatrix} A\alpha + B\beta \\ \gamma \end{bmatrix} = 0.$$

If $A\alpha + B\beta \neq 0$, there exists a $(j-1)$-constant direction, which is a contradiction.
So $A\alpha + B\beta = 0$. Then $\gamma \in N[B'_{(j-1)} \ R_{(j-1)}]'$ and so by Remark 3.1.2, $\gamma \in N[C_{(j-1)}]$.
From the middle block row of (6.5), we have then

$$0 = B'Q_{(j-1)}[A\alpha + B\beta] + C'\alpha + R\beta + B'C_{(j-1)}\gamma = C'\alpha + R\beta.$$

This shows that $[\alpha' \ \beta']' \in N[\Lambda]$, again a contradiction.

We have now completed the program set out at the start of this section, of showing
that by successive removals of superfluous controls and 1-constant directions,
ultimately all j-constant directions, for any j, are eliminated. Motivated by
this result, it is evidently sensible to make the following definition.

Definition: The optimal control problem is called singular whenever Λ is
singular. Otherwise, it is called nonsingular.

The procedure for solving a singular problem can be outlined as follows:

1. Determine the nullity of Λ, say s. If $[B' \ R]'$ has full rank proceed to 2.
    If not, eliminate any unnecessary controls by the procedure of Section 4.

2. Let $\ell = \dim I_1$. If $\ell = 0$, we have a nonsingular problem. If $\ell > 0$, reduce
    the state dimension by ℓ as described in Section 5. If $\ell = n$, we have a zero
    state dimension problem. If $\ell < n$, return to 1.

3. Cycle through 1 and 2 until the procedure terminates. This is guaranteed by the
    above theory, and moreover, it is guaranteed that all of the constant directions
    of the original problem are determined in at most n applications of 1 and 2.

4. Determine the solution of the terminating control problem and trace back through
    the reduction procedure to construct the solution of the originally given problem.

In the commonly occurring case of [A, B] completely reachable, we cannot terminate with a zero control dimension problem. This follows from the fact that complete
reachability is preserved under the reduction procedure.

TIME-VARYING PROBLEMS, MISCELLANEOUS POINTS AND SUMMARIZING REMARKS

In Chapter III, results for the time-varying linear-quadratic control problem
e presented in detail for the continuous-time case. However, for the derivation of
ese results, certain constancy of rank assumptions are required on the interval of
terest. These assumptions might be thought of as a constant structure requirement.

A similar idea applies to the extension of the constant coefficient results
ained in Sections 2 - 6 to the time-varying case. For example, suppose that we
sider the interval $[M, N]$; we can define the matrix $\Lambda(i)$ for each $i = M,...,N-1$.
s(i) be the nullity of $\Lambda(i)$, let $W(i)$ be a basis matrix for $N(\Lambda(i))$ and
$\ell(i)$ be the rank of $W_1(i)$, with $W_1(i)$ the first n rows of $W(i)$. Then
) and $\ell(i)$ must be constant on the interval $[M, N-1]$ for the reduction of
control and state dimensions previously described to be valid. Essentially, for
s reason we have for the time-varying case the following definition.

Definition: Suppose $1 \leq j \leq N-M-1$. The n-vector $\alpha(i)$ is called a
egenerate direction at i of (2.5) on the interval $[M, N]$ if and only if
, S)$\alpha(i)$ is the same function of $A(\cdot)$, $B(\cdot)$, $Q(\cdot)$, $C(\cdot)$ and $R(\cdot)$ for all
i $\leq N-j$ and all S for which (2.5) has a solution.

For a number of reasons we have chosen not to set out the time-varying results
detail. First, the notation becomes complex and, save for the idea of constancy
structure, no principles other than those already considered for the time-invariant
e are needed. Second, the principle that it can be done has already been estab-
hed in [3], though only for the special case of scalar covariances. Perhaps it is
thwhile to point out at this stage that the constancy of structure requirement in
appears via the condition that the covariances possess what is termed a definite
ative order. Third, one of the major reasons for studying constant directions and
associated reduction in order of the Riccati equation is the computational saving.
derivations in this paper require basis changes of both the state and control
ces. For the time-varying case, these basis changes need to be carried out for each
M, ..., N-1 so for N-M large and no functional relation between the coefficient
rices at each time instant, it is not unlikely that the hoped for computational
antage could be nullified.

er Miscellaneous Points

1. If there are any constant directions at all, there must be 1-constant
ections. Therefore, the testing for existence of any constant directions is easily
cuted by looking at the matrix Λ.

2. Let α be any constant direction, and let Π_1, Π_2 be any two constant
tions of the Riccati equation (2.5) with $B'\Pi_i B + R \geq 0$ and $N[B'\Pi_i B + R]$
$[A'\Pi_i B + C]$. Since $P(N-j, \Pi_i) = \Pi_i$ for all j and $i = 1, 2$, one has $\Pi_1\alpha = \Pi_2\alpha$.

3. In case R > 0, C = 0 (which is a common situation in regulator and filtering theory), and if also A is nonsingular, there are no constant directions. (This follows easily from an examination of Λ). On the other hand, R > 0, C = 0 and A singular implies that any vector in the nullspace of A is a 1-constant direction. (This is easy to see intuitively; use of the zero control ensures that the next state is zero).

4. The condition that $N(\Lambda)$ be nonempty is equivalent to the condition that $R + C'(zI-A)^{-1}B$ have a transmission zero at z = 0, [8]. If A is nonsingular, this is equivalent to demanding that the matrix $R + C'(zI-A)^{-1}B$ be singular at z = 0. In turn this is equivalent to $R + C'(zI-A)^{-1}B + B'(z^{-1}I-A)^{-1}C$ $+ B'(z^{-1}I-A)^{-1}Q(zI-A)^{-1}B$ having this property.

5. In one interesting case, all directions are j-constant directions for some j. Let [A, B] be completely reachable, suppose the optimal control problem has a solution on [0, 2n] for some S, and suppose that $R + C'(zI-A)^{-1}B + B'(z^{-1}I-A')^{-1}C$ $+ B'(z^{-1}I-A')^{-1}Q(zI-A)^{-1}B = 0$ for all z not an eigenvalue, or inverse of an eigenvalue, of A. To see that all directions are constant directions, let x(n) be arbitrary and fixed, x(0) = 0. Let U_0^{n-1} be a control sequence taking x(0) = 0 to the prescribed x(n) and let U_n^{2n-1} be for the moment arbitrary. Because the control problem has a solution on [0, 2n], it follows that

$$0 \le V_{2n}[0, U_0^{2n-1}, S] = V_n[0, U_0^{n-1}, 0] + V_n[x(n), U_n^{2n-1}, S].$$

[If the inequality failed, V_{2n} could be made as negative as desired by scaling U_0^{2n-1}]. Now suppose that U_n^{2n-1} is such that x(2n) = 0. Define u(k) = 0 for k ∉ [0, 2n-1]. Use Parseval's theorem to evaluate

$$\sum_{-\infty}^{+\infty} [x'(k)Qx(k) + 2x'(k)Cu(k) + u'(k)Ru(k)] = V_{2n}[0, U_0^{2n-1}, S].$$

The frequency domain equality then yields $V_{2n}[0, U_0^{2n-1}, S] = 0$. (A similar argument has been used in the continuous-time linear-quadratic problems in the proof of Theorem 4 of [9]). Hence $V_n[x(n), U_n^{2n-1}, S]$ attains its lower bound, via., $-V_n[0, U_0^{n-1}, 0]$ Any U_n^{2n-1} causing x(2n) = 0 is optimal, and since x(n) is arbitrary, all directions are constant. This means incidentally that the transient solution of (2.5) will agree with the steady state solution after at most n steps. Finally, note that it is not necessarily the case that if all directions are constant, the frequency domain relation holds.

6. The matrix Λ is square, so that if $N(\Lambda)$ is nonempty, so is $N(\Lambda')$, and the question then arises as to what significance, if any, attaches to vectors in $N(\Lambda')$ One can make the trivial observation that nonzero vectors in $N(\Lambda')$ are associated with 1-constant directions for a dual problem [where x(i+1) = A'x(i) + Cu(i) and C in the performance index is replaced by B], but beyond this, not much can be said. In particular, it does not seem to be possible to make statements relevant to the

mal control problem.

7. Suppose that for some j, α, S and S with $S > S_0$, one has $V_j^*[\alpha, S]$
$_j^*[\alpha, S_0]$. Thus α is a j-constant direction. It is immediate that the control
uence minimizing $V_j[\alpha, S]$ also minimizes $V_j[\alpha, S_0]$. The point of this remark
that a control sequence minimizing $V_j[\alpha, S_0]$ need not minimize $V_j[\alpha, S]$; this
shown by the example below. On the other hand, if $V_j^*[\alpha, S_0 - \eta I]$ exists for some
0, the control sequence minimizing $V_j[\alpha, S_0]$ must carry α optimally to zero,
Lemma 3.2, and such a control sequence also minimizes $V_j[\alpha, S]$. For the example,
e $A = B = C = R = 1$, $Q = 0$. Then $V_1[x(N-1), u(N-1), S]$
$S+1)u^2(N-1) + 2(S+1)u(N-1)x(N-1) + x^2(N-1)$; if $S > -1$, $u(N-1) = -x(N-1)$ is the
que optimal control, while if $S = -1$, any value of $u(N-1)$ is optimal. The
imum norm value is of course zero, and this is certainly not optimal for $S > -1$
$x(N-1) \neq 0$.

8. For time-invariant linear-quadratic problems, the so-called Chandrasekhar
orithms [5, 6] appear very attractive computationally. It would therefore be
eresting to connect the ideas of this chapter, at the computational level, with
Chandrasekhar algorithms. Conceptually, it is fairly clear that a connection
ald be fruitful. This is because the Chandrasekhar algorithms work with first
ferences $P(i+1, S) - P(i, S)$ of Riccati equation solutions, and are advantaged
hese quantities having low rank; the more linearly independent constant directions
e are, the lower the rank of the first difference. At a more detailed level, some
would however be required in separating the control dimension reduction step from
state dimension reduction step, and one would need to vary the proofs of existence
ertain orthogonalizing transformations in [6] to cope with various possible singul-
ies. Finally, extensions to the time-varying case could pose problems.

9. In [4], a "structure algorithm" is presented which bears on linear quadratic
lems in which $S \geq 0$ and $\begin{bmatrix} Q & C \\ C & R \end{bmatrix} \geq 0$. The manipulations for eliminating super-
us controls, characterizing and eliminating 1-constant directions, and identifying
with states which can be taken to zero optimally in one step are equivalent to
pulations in [4]. The full extent of the parallels, and the point at which they
k down because of the nonnegativity requirement of [4] has yet to be explained.

10. Though we have confined our discussions to control problems, we could equally
have worked within a framework of filtering and covariance factorization, as in
Combination of the ideas here and of [3] will readily yield the results.

arizing Remarks

There have been two main themes of this chapter. First, we have discussed pro-
ies of constant directions in the context of the most general linear-quadratic
rol problem. The most important constant directions, viz. 1-constant directions,
characterized via the null-space of a certain simply constructed matrix, and all

constant directions are characterized in terms of optimal controls yielding trajectories which terminate in the zero state. Second, we have shown how the existence of constant directions may be exploited in solving an optimal control problem. They may be eliminated to yield a lower dimension problem, the solution of which determines the solution of the original problem, with the adjunction of certain quantities computed in the construction of the lower dimension problem.

APPENDIX IV.A

DEFINITIONS OF COEFFICIENT MATRICES

The matrices used in equation (5.8) are defined as follows:

$$\hat{A} = A_{11}$$

$$\hat{B} = [\hat{B}_1 \quad \hat{B}_2] = [B_{12} \quad B_{11}]$$

$$\hat{Q} = A_{11}'(Q_{12} + C_{12}Z)A_{21} + A_{21}'(Q_{12} + C_{12}Z)'A_{11}$$
$$+ A_{21}'(Q_{22} + C_{22}Z)A_{21} + Q_{11}$$

$$\hat{C} = [\hat{C}_1 \quad \hat{C}_2]$$

$$\hat{C}_1 = A_{11}'(Q_{12} + C_{12}Z)B_{22} + A_{21}'(Q_{12} + C_{12}Z)'B_{12}$$
$$+ A_{21}'(Q_{22} + C_{22}Z)B_{22} + C_{12}$$

$$\hat{C}_2 = A_{11}'(Q_{12} + C_{12}Z)B_{21} + A_{21}'(Q_{12} + C_{12}Z)'B_{11}$$
$$+ A_{21}'(Q_{22} + C_{22}Z)B_{21} + C_{11}$$

$$\hat{R} = \begin{bmatrix} \hat{R}_{11} & \hat{R}_{12} \\ \hat{R}_{12}' & R_{22} \end{bmatrix}$$

$$\hat{R}_{11} = B_{12}'(Q_{12} + C_{12}Z)B_{22} + B_{22}'(Q_{12} + C_{12}Z)'B_{12}$$
$$+ B_{22}'(Q_{22} + C_{22}Z)B_{22} + R_{22}$$

$$\hat{R}_{12} = B_{12}'(Q_{12} + C_{12}Z)B_{21} + B_{22}'(Q_{12} + C_{12}Z)'B_{11}$$
$$+ B_{22}'(Q_{22} + C_{22}Z)B_{21} + R_{12}'$$

$$\hat{R}_{22} = B_{11}'(Q_{12} + C_{12}Z)B_{21} + B_{21}'(Q_{12} + C_{12}Z)'B_{11}$$
$$+ B_{21}'(Q_{22} + C_{22}Z)B_{21} + R_{11}$$

ERENCES

D. Rappaport, "Constant directions of the Riccati equation", *Automatica*, Vol. 8, 1972, pp. 175-186.

R.S. Bucy, D. Rappaport and L.M. Silverman, "Correlated noise filtering and invariant directions of the Riccati equation", *IEEE Trans. Automatic Control*, Vol. AC-15, 1970, pp. 535-540.

M. Gevers and T. Kailath, "Constant, predictable and degenerate directions of the discrete-time Riccati equation", *Automatica*, Vol. 9, 1973, pp. 699-711.

L.M. Silverman, "Discrete Riccati Equations", Control and Dynamic Systems, Vol. 12, ed. C.T. Leondes, Academic Press, N.Y., 1976, pp. 313-386.

M. Morf, G.S. Sidhu and T. Kailath, "Some new algorithms for recursive estim-ation in constant, linear, discrete-time systems", *IEEE Trans. Automatic Control*, Vol. AC-19, 1974, pp. 315-323.

M. Morf and T. Kailath, "Square-root algorithms for least squares estimation", *IEEE Trans. Automatic Control*, Vol. AC-20, 1975, pp. 487-497.

D.J. Clements and B.D.O. Anderson, "Linear-quadratic discrete-time control and constant directions", *Automatica*, Vol. 13, 1977, to appear.

H.H. Rosenbrock, State Space and Multivariable Theory, T. Nelson and Sons, London, 1970.

J.C. Willems, "Least squares stationary optimal control and the algebraic Riccati equation", *IEEE Trans. Automatic Control*, Vol. AC-16, 1971, pp. 621-634.

CHAPTER V

OPEN QUESTIONS

There are a number of open questions arising from the work discussed in the previous chapters. Some problems have already been mentioned in those chapters; we will not repeat these here.

Perhaps one of the more obvious questions is: how can one cope with a problem for which there are a finite number of structural changes along the interval of interest? (Trying to cope with an infinite number of structural changes seems impossibly difficult). Structural changes correspond to the matrix P^* possessing jumps at the points of structural change, and the problem is to determine the magnitude of the jumps in the $P^*(\cdot)$ matrix at the points of structural change or, equivalently, to match the $P^*(\cdot)$ matrices on the individual intervals at the junction points.

A question which is clearly related is that of joining up controls and trajectorie between singular and nonsingular regimes, or between dissimilar singular regimes. Ther are both quantitative issues and qualitative issues involved; some results are surveye in [1].

Recently "high order" maximum principles [2] have been applied to singular problems. It would be of interest to closely relate the methods of Chapter III to the results achievable by the high order maximum principle; to the extent that the Euler-Lagrange equations are likely to figure prominently, recent work on their use in singular problems [3] is likely to be relevant.

The possibility of reduction in computational complexity of a general singular discrete-time problem has been demonstrated in the previous chapter. For particular classes of problems there are a number of efficient computational algorithms, e.g. the square root filtering algorithm and the Chandrasekhar algorithm mentioned in Section 7 of Chapter IV. It would be of interest to set up the connections between these algorithms and our own.

Finally, we draw attention to what is almost a problem of logic. Often a singular linear-quadratic control problem (or nonnegativity problem) is the result of linearization of a nonlinear system about a nominal control and trajectory. It is therefore clear that there are, in effect, implied constraints on both the controls and states in the second variation linear-quadratic problem. This would obviously preclude the appearance of delta functions in an optimal control problem (delta functions being the limit of continuous functions with upper bound approaching infinity). We must then ask to what extent the calculation of singular controls is a valid exercise. Of course, in a specific problem, regularization of any singular second variation problem [though the addition of a term $\varepsilon u'(t)u(t)$ to the loss function for some $\varepsilon > 0$] will determine control gains which are useable to give a valid approximation to the adjustment in the optimal control stemming from a small enough adjustment in the initial state.

One approach to solving singular control problems is to regularize them, i.e. to a term $\varepsilon u' u$ with $\varepsilon > 0$ and small to the loss function, thereby obtaining a singular problem with a solution in some way close to that of the singular problem. resulting nonsingular problem - a "cheap control" problem - is normally numeric- y ill-conditioned, and special approaches are being developed to solve such prob- s, see e.g. [4]. It would be of interest to check whether the reduction procedures posed for singular problems could be profitably used also on nonsingular, cheap trol, problems.

A further problem is to tidy up some of the results presented here for problems h end-point constraints. The robustness results for the case when the final state partially but not completely constrained have not been fully developed. An algor- n for constructing optimal controls and the optimal performance index has not yet a given, but almost certainly, it should be straightforward to obtain as an exten- n of the free end-point algorithm.

Another area left untouched relates to allowing semi-infinite intervals $[t_0, \infty)$ tead of the finite intervals $[t_0, t_f]$ considered throughout this book. There is course a fairly extensive theory, see e.g. [5], for time-invariant nonsingular blems and for the time-varying linear regulator problem applicable to semi-infinite ervals, and much constitutes non-trivial extension of the finite interval results. this suggests that it might be fruitful to study the general singular problem on semi-infinite interval.

RENCES

D.J. Bell and D.H. Jacobson, Singular Optimal Control Problems, Academic Press, New York, 1975.

A.J. Krener, "The high order Maximal Principle and its application to singular extremals", SIAM J. Contr. Opt., Vol. 15, 1977, pp. 256-293.

S.L. Campbell, "Optimal control of autonomous linear processes with singular matrices in the quadratic cost functional", SIAM J. Contr. Opt., Vol. 14, 1976, pp. 1092-1106.

R.E. O'Malley, "A more direct solution of the nearly singular linear regulator problem", SIAM J. Contr. Opt., Vol. 14, 1976, pp. 1063-1077.

B.D.O. Anderson and J.B. Moore, Linear Optimal Control, Prentice Hall, New Jersey, 1971.